Advanced Structured Materials

Volume 35

For further volumes:
http://www.springer.com/series/8611

Nicola Bonora · Eric Brown
Editors

Numerical Modeling of Materials Under Extreme Conditions

Springer

Editors
Nicola Bonora
DICEM - Dipartimento di Ingegneria Civile e Meccanica
University of Cassino and Southern Lazio
Cassino
Italy

Eric Brown
Los Alamos National Laboratory
Los Alamos
USA

ISSN 1869-8433 ISSN 1869-8441 (electronic)
ISBN 978-3-642-54257-2 ISBN 978-3-642-54258-9 (eBook)
DOI 10.1007/978-3-642-54258-9
Springer Heidelberg New York Dordrecht London

Library of Congress Control Number: 2014939033

Printed on acid-free paper

Springer is part of Springer Science+Business Media (www.springer.com)

Contents

Particulate Composites Under High Strain Rate and Shock Loading

J. L. Jordan and E. B. Herbold

Abstract Polymer-matrix particulate composites consist of individual particles of more than one material dispersed throughout and held together by a polymer binder. The mechanical and physical properties of the composite depend on the mechanical and physical properties of the individual components, particularly the polymer binder; their loading density; the shape and size of the particles; the interfacial adhesion; residual stresses; and matrix porosity. Systematic studies of the effects of volume fraction and microstructure on the behavior of these polymer-based composites are critical. The behavior of polymer-matrix particulate composites at intermediate to high strain rates has not been investigated in detail in the literature. The testing strain rate can greatly affect the behavior of these composites due to the dependency on rate dependant phase changes in the polymer binder. The intermediate strain rate behavior ($\sim 10^3$–10^4/s) is studied using a split Hopkinson pressure bar. Shock, or high strain rate, properties of these composite materials have been investigated using gas gun and explosive loading techniques. This chapter will review results from recent experimental studies on the properties of polymer-based particulate composites containing metal and metal oxide powders.

Keywords Particulate composite · High strain rate · Shock loading · Hugoniot

J. L. Jordan (✉)
Air Force Office of Scientific Research, Arlington, VA 22203, USA
e-mail: jennifer.jordan.6@us.af.mil

E. B. Herbold
Lawrence Livermore National Laboratory, Livermore, CA 94550, USA
e-mail: herbold1@llnl.gov

N. Bonora and E. Brown (eds.), *Numerical Modeling of Materials Under Extreme Conditions*, Advanced Structured Materials 35, DOI: 10.1007/978-3-642-54258-9_1,

1 Introduction

Polymer-based particulate composites are comprised of one or more metallic, ceramic, or polymeric particulates in a polymer binder. The elastic and quasi-static properties of particulate composites have been well studied, both experimentally and theoretically [1–4]. However, the dynamic and shock properties of these materials, particularly when more than one particulate phase are present in the composite.

There are several types of polymer-based particulate composites. One system prevalent in the literature is based on a glassy epoxy binder. There is considerable data on composites based on this type of binder, and these composites will be the focus of this review. Particulate composites have complex microstructures consisting of stiffer particles in a softer matrix, as shown in Fig. 1, which is a representative microstructure of aluminum–tungsten-epoxy. These composites are locally very heterogeneous, with contact between particles, particle chains, and particle clustering.

In this review, the methods for dynamic strength and shock properties in particulate composites will be discussed. The literature for the high strain rate deformation of particulate composites, primarily epoxy-based, will be reviewed. Several key features and analyses for these composites will be discussed, namely particle deformation and mechanical percolation thresholds. Finally, the available literature on the shock response of epoxy-based composites will be discussed and summarized.

2 Experimental Techniques to Characterize Particulate Composites

2.1 Dynamic Mechanical Characterization

A traditional method for determining the dynamic stress–strain response of materials is the Split Hopkinson Pressure Bar (SHPB) [5], shown schematically in Fig. 2. In this experiment, a striker bar is impacted into one end of a long input rod introducing a stress wave into that rod. The wave travels down the bar to the specimen, where a portion of the wave is reflected and a portion is transmitted. The waves are measured using strain gauges typically mounted halfway down the rods. Using the measured strains, the stress–strain behavior of the sample can be determined [5].

Care must be taken when testing polymer-based particulate composites with a split Hopkinson pressure bar. These materials generally have low impedance (product of sound speed and density), so metallic bars with similar impedance should be chosen to allow for transmission of the stress wave, rather than seeing the sample as a free surface. Typically, for the epoxy-based composites, the use of aluminum input and output bars are sufficient to provide measurable strain signals. Additionally, the use of semiconductor strain gages, which have a gage factor

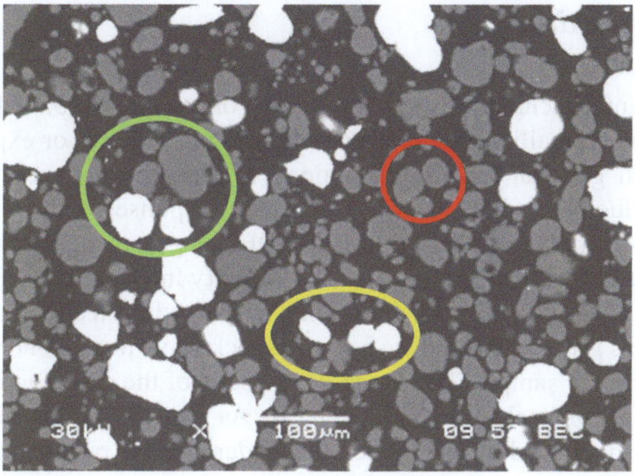

Fig. 1 Micrograph of aluminum–tungsten-epoxy composite, where the aluminum are *light gray particles* and the tungsten are *white particles*. The *circles* indicate contact between particles, particle chains, and particle clustering

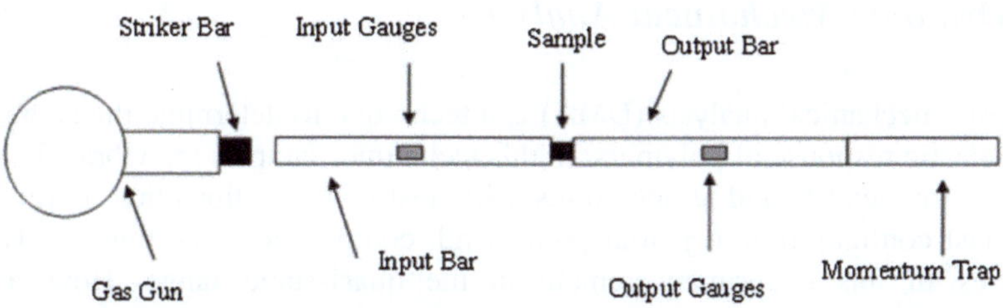

Fig. 2 Schematic of a split Hopkinson pressure bar (SHPB)

of ~140 versus a traditional foil gage with ~2–4, can be used to increase the voltage signals [6]. Some authors have found that the use of pulse shaper can improve the test configuration by inducing a longer rise time and decreasing wave dispersion [7].

Taylor rod-on-anvil impact experiments can be used to access the dynamic deformation of particulate composites, in fact most materials, in the strain rate regime of 10^3–10^4 s^{-1}. In these experiments, a rod of material is accelerated using a gas gun, or similar instrument, and impacted into a rigid anvil. The deformation of the specimen gives insight into its dynamic mechanical properties. In traditional Taylor impact experiments on metals [8], the samples were recovered and measurements of the deformation were made on the samples. As the test technique has been applied to polymers and particulate composites, new analysis [9] and in situ photographic techniques [10–14] have been required to understand the dynamic deformation of the complex materials.

2.2 Shock Loading

The experimental techniques used for shock compression [15, 16] of particulate composites are typically accomplished with gas gun [17, 18] or explosive loading [17, 19]. With gas gun loading techniques, a flyer plate impacts the sample of interest introducing a shock wave. The shock response of the material can be determined using polyvinylidene floride (PVDF) gauges, manganin gauges [17, 20], electromagnetic (EM) gauges [17, 21], and Velocity Interferometry System for Any Reflector (VISAR) [18, 22–24]. In an explosive loading technique, an explosive plane wave lens is used to introduce a planar shock wave into a donor plate, which is in contract with the sample. The shock wave speed of the donor and sample can be measured using piezoelectric pins [17, 19]. As this chapter is not a review of experimental shock loading techniques, the reader is left to pursue the references for further details.

2.3 Mechanical Properties of Particulate Composites Dynamic Mechanical Analysis

Dynamic mechanical analysis (DMA) is a technique to determine the small strain visco-elastic response of polymers. In this technique, samples are vibrated across a range of frequencies and temperatures using test configurations that mimic quasi-static test configuration, e.g. four point bend, compression, and tension. The frequencies in this experiment remain in the quasi-static range. However, the experimental results can be used to predict the high strain rate behavior and phase changes of polymers [25, 26] and particulate composites [27, 28]. As expected, the addition of increasing volume fraction of particulate results in an increase in the measured room temperature modulus. However, the phase transitions in the polymer binder of particulate composites containing micron-sized particles have been shown to be unaffected by the presence of the filler.

3 High Strain Rate Properties of Particulate Composites

The behavior of epoxy-based particulate composites under high strain rate loading is highly dependent on the volume fraction and material properties of the filler. Alumina-epoxy composites show deformation and fracture under high rate loading [7], where aluminum-epoxy composites show "plastic" deformation with an intact sample, although they will exhibit some radial cracking. Figure 3a and b shows an aluminum-epoxy composite before and after deformation, where the deformation of the aluminum with the binder can be seen. Although the alumina-epoxy composites are significantly stronger than aluminum-epoxy composites with the same

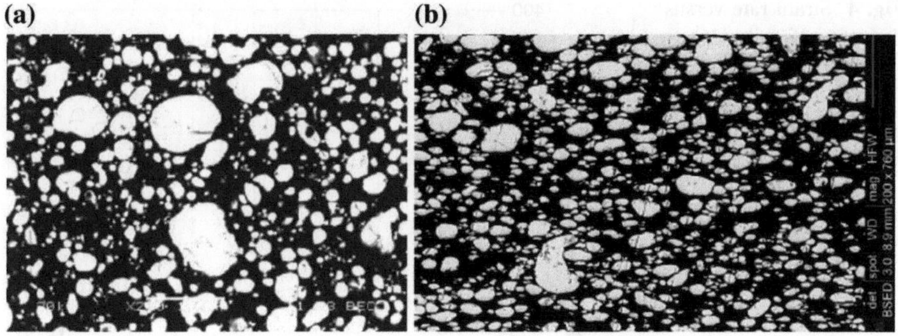

Fig. 3 Aluminum-epoxy composite **a** before, and **b** after deformation

volume fraction of filler, as shown in Fig. 4, the alumina particles are not believed to be deforming with the binder resulting in fracture of the composite under high rate loading.

For composites containing more than one particulate phase, the dependence of the compressive flow stress on the strain rate can be more complex. In nickel–aluminum composites with varying amounts of nickel and aluminum, the flow stress exhibited dependence on aluminum volume fraction, nickel volume fraction, and aluminum particle size [29]. The strength of the materials with lower total volume fractions of particles is either higher or approximately the same as materials with higher volume fractions [29].

The dynamic tensile properties of these composites are not typically measured due to the low strain to failure and complex experimental set-up required. However, Owens and Tippur [30] have developed a split Hopkison tensile bar (SHTB) ideally suited to testing of epoxy-based composites. They examined the effect of volume fraction and particle size on the high rate tensile behavior of glass sphere-epoxy composites. In these composites, increasing the volume fraction of glass microspheres resulted in a reduced failure stress, most likely due to the increased interface surface area allowing more sites for interfacial debonding. Above a certain minimum particle size, increasing the particle size also reduced the strength of the composites.

Taylor impact testing of polymer-based particulate composite materials is still in its infancy. These materials have required the development of both elastic-plastic analysis [9] and high-speed cameras to capture in situ deformation. Taylor impact experiments on a series of nickel–aluminum-epoxy composites have shown that the critical velocity and dynamic yield stress both decrease with increased volume fractions of particles [14]. The particle size of the aluminum phase in nickel–aluminum-epoxy and aluminum–iron oxide-epoxy is shown to greatly affect the mechanical properties [12–14]. With micron sized aluminum particles, decreasing the particle size increased the critical velocity [14]. The critical velocity in these composites showed a complex behavior based on particle size, where below a percolation threshold, the smaller particles showed lower dynamic

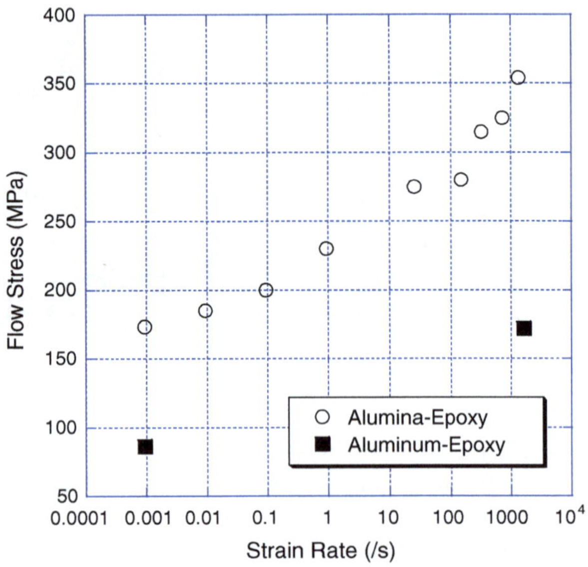

Fig. 4 Strain rate versus flow stress for alumina-epoxy [7] and aluminum-epoxy [29]

yield stress and above the percolation threshold the reverse held true [14]. For composites containing nano-sized aluminum [12, 13], the high strain rate behavior transitioned to fracture, with the nano-sized aluminum containing composites exhibiting higher strength than similar micron-sized composites. This behavior was attributed to the nano-sized particles causing increased polymer chain entanglements resulting in increased strength but decreased ductility [12, 13].

4 Particle Deformation

Particles have been used to measure internal strains since the late 1970s [31–34]. Typically, aluminum, or other metallic inclusions, have been embedded in fiber-reinforced composites and the strains measured using X-Ray Diffraction (XRD) or Micro Raman Spectroscopy (MRS) [32–34]. However, these particles have been an addition to the material being studied rather than the material itself. Additionally, glass bead composites used to simulate geomaterials have been recovered and used as internal measures of strain [35].

Recently, the metallic particles, particularly aluminum, present in epoxy-based particulate composites have been used as post-mortem strain gauges [29]. In this technique, the deformation of several (at least 100) particles is measured to determine the average strain in the particulate phase. Assuming that the particle volume is conserved during the deformation and that a spheroidal particle with an initial aspect ratio of 1 will deform as an oblate spheroid with its minor axis along the deformation direction, then the particle strain can be related to the particle aspect ratio [29].

The technique of using the particles themselves as post-mortem strain gauges can reveal much about the stress transfer in particulate composites. As seen in Fig. 5, the strain in aluminum particles is consistently higher than the global strain in the composite, indicating efficient load transfer between the epoxy and the stiffer aluminum inclusions [29]. In numerical simulations comparing clustered and unclustered microstructures, local stresses were higher in the clustered microstructures resulting in increased damage by particle fracture and matrix-particle decohesion [36]. Similarly, experimental results of metal-epoxy particulate composites have shown that aluminum particles close to a harder second phase, e.g. nickel or tungsten particles, show increased deformation over aluminum particles far from the second phase, as shown in Fig. 6 [29, 37].

5 Mechanical Percolation Thresholds

Percolation thresholds are typically considered in composites containing a conducting phase embedded in a non-conducting matrix [38]. Above a critical concentration in these materials, i.e. the percolation threshold, the conducting phase forms a three-dimensional network resulting in a continuous conductive path across the composite. Mechanical percolation thresholds have been observed in consolidated metal powder compacts containing a harder second inclusion, e.g. steel spheres in an aluminum matrix [39]. In these composites, the inclusion network will support the applied compaction pressure when a continuous network is present resulting in a lower relative density of the compact.

Lekatou et al. [40] found that glass beads in an epoxy matrix showed an increase in quasi-static tensile modulus above a percolation threshold due to the cage-like structure that was formed. Although the tensile strength of particulate composites typically decreases with increasing filler amount, the tensile strength has been observed to decrease dramatically at the electrical percolation threshold [41], which may be attributed to the formation of continuous networks that allow for crack propagation as well as electrical conductivity [40–42]. An additional threshold may exist at very high filler concentrations resulting from the interconnectedness of the polymer matrix below a certain filler concentration [43].

In the high strain rate regimes particularly applicable for this article, White et al. [14] observed percolation thresholds in aluminum-epoxy and aluminum–nickel-epoxy composites in Taylor experiments, where the dynamic yield stress showed a dramatic decrease at approximately 0.35 volume fraction particles. As the particles form a three-dimensional network, the stress transfer in these materials may become inefficient, i.e. the matrix is not transferring the load equally across the composite.

Similarly, in aluminum-epoxy composites tested at high strain rates using a split Hopkinson pressure bar, there appears to be a threshold above which increased

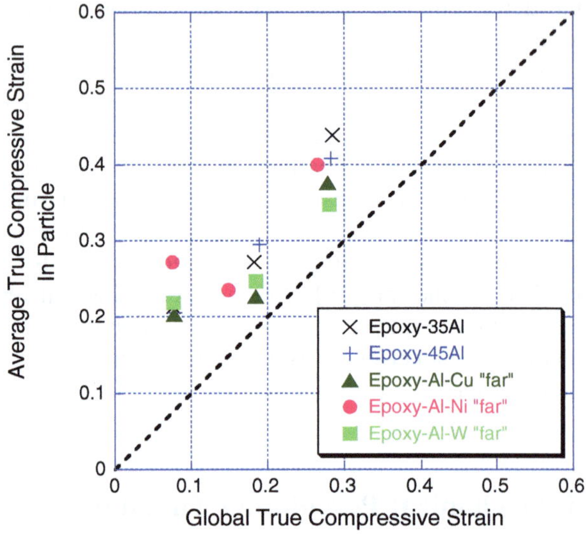

Fig. 5 Quasi-static global true compressive strain versus average true compressive strain for single phase (Al only) particulate composites and two-phase particulate composites with Al particles far from the second phase [29]

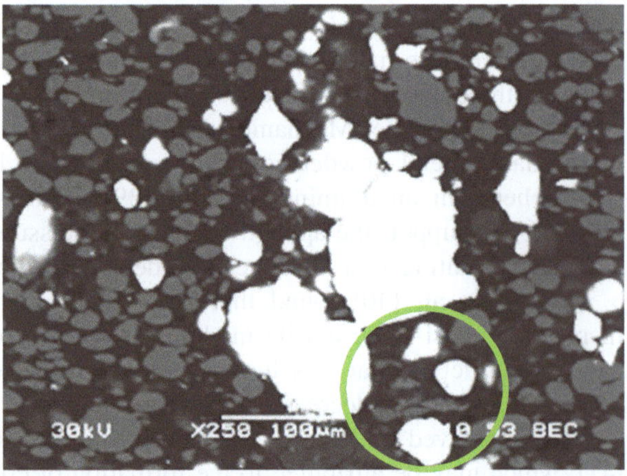

Fig. 6 Aluminum–tungsten-epoxy showing increased deformation of aluminum particles in close contact with tungsten particles

volume fraction of particles results in decreased peak stress, as shown in Fig. 7a. A threshold such as this typically indicates the interaction of particles, which may lead to inefficient stress transfer across the entire composite and reduced strength. The particles in these composites were measured post-mortem to determine the strain in the aluminum. The epoxy-13Al sample containing 13 vol.% aluminum showed significantly less strain in the aluminum particles than in the bulk composite. This suggests that, although the addition of 13 vol.% aluminum results in increased flow stress in the composite, the epoxy is absorbing more of the deformation. It may be that particle interaction is required for the aluminum particles to exhibit the same amount of deformation as seen in the bulk sample.

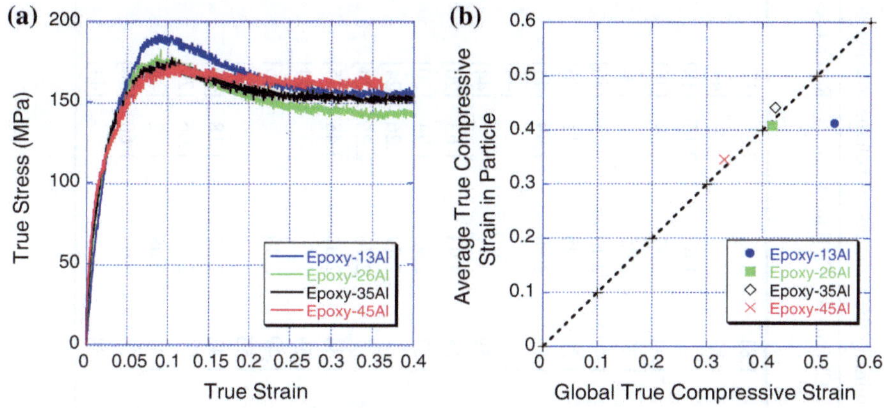

Fig. 7 **a** Stress versus strain, and **b** global true compressive strain versus the average compressive strain in the particles for aluminum-epoxy composites with increasing amounts of aluminum. The volume fraction of aluminum is indicated by the number in the sample label, i.e. Epoxy-13Al has 13 vol.% aluminum

6 Shock Loading of Particulate Composites

The shock loading of alumina-epoxy has been extensively studied in the literature [20, 22–24, 44]. Additionally, other single particulate phase [18] and multi-particulate composites [17, 19, 45] have been studied. A summary of the epoxy-based particulate composite materials characterized under shock loading is presented in Table 1.

In the pure epoxy binder [23, 47–49], Carter and Marsh [47] observed that a linear fit to the shock velocity—particle velocity curve does not extrapolate to the bulk sound speed in the material, which was attributed to the distance between polymer chains compressing while leaving the polymer backbones intact. Additionally, a high-pressure phase transition appears to be present at ~23 GPa, which may be due to interchain chemical reactions [47]. Variation of composition of similar epoxy systems appears to have little effect on the Hugoniot behavior in the measured pressure regimes [23, 47–49].

The addition of particulate phases into the epoxy can result in a complex shock response, as shown by the extensive literature on alumina-epoxy composites [20, 23, 24, 44, 49–53]. Compared with the epoxy binder, the alumina-epoxy composite exhibits a decreased compressibility, with increasing volume fraction of alumina further decreasing the compressibility [20, 23]. In the composites, the propagated waves have been observed as rounded in certain pressure regimes [44, 48]. At high input stresses, the composites exhibit a viscoelastic response, which transitions to a more complex response at low input stresses due to the temporal dynamics of the viscous mechanisms [24]. Decreasing the particle size appears to decrease the effect of viscous processes in the composites as shown by the steeper wave profiles [44]. The release wave has been experimentally determined to

Table 1 Epoxy-based particulate composites studied under shocked loading, where the shaded rows have nominally the same epoxy volume fraction

Material	Volume fraction									Densitv (g/cm^3)	C_1 (km/s)	C_s (km/s)	C_0 (km/s)	S	References
	Epoxy	Al_2O_3	A1	Fe_2O_3	WC	MnO_2	Cu	Ni	W						
Al-MnO_2	0.45									2.598	3.36	1.81	2.63	2.86	[17]
All	0.52	0.48								2.496	3.31	1.82			[44]
WCE	0.54				0.45					7.77			1.65	2.15	[18]
Epoxy-45Al	0.55		0.45							1.875	3.13	1.61	2.52	1.89	[46]
Epoxy-Al-Cu	0.55		0.35				0.1			2.475	2.77	1.40	2.25		[46]
Epoxy-Al-Ni	0.55		0.35					0.1		2.513	2.95	1.49	2.40		[46]
Epoxy-A1-W	0.55		0.35						0.1	3.652	2.39	1.20	1.95	2.11	[46]
A12	0.57	0.43								2.377	3.20	1.72	2.88	1.99	[44]
A13	0.57	0.43								2.389	3.13	1.70			[44]
A14	0.57	0.43								2.391	3.03	1.59			[44]
Fullv loaded	0.57	0.43								2.28	3.16	1.68	2.93	1.63	[20]
300/100/20	0.58	0.42								2.37	3.40	1.57	2.93	1.94	[50]
F1	0.58	0.42								2.376	3.13		2.79	2.26	[24]
F2	0.58	0.42								2.429	3.26		2.76	2.26	[24]
Al-Fe_2O_3	0.60		0.16	0.24						2.366	2.94	1.34	2.02	3.48	[45]
A15	0.62	0.38								2.233	3.03	1.59			[44]
Epoxy-35Al	0.65		0.35							1.725	2.95	1.49	2.40		[46]
A16	0.66	0.34								2.377	2.99	1.56			[44]
200/100/20	0.66	0.34								2.13	3.09	1.43	2.87	1.18	[50]
Half loaded	0.75	0.25								1.87	2.78	14	2.63	1.66	[20]
Al-Fe_2O_3	0.78		0.09	0.13						2.047	2.62	1.31	3.08	1.22	[45]
100/100/20	0.80	0.20								1.76	2.82	1.23	2.66	1.60	[50]
A17	0.80	0.20								2.121	2.74	1.34			[44]

The material designation originates from the original reference

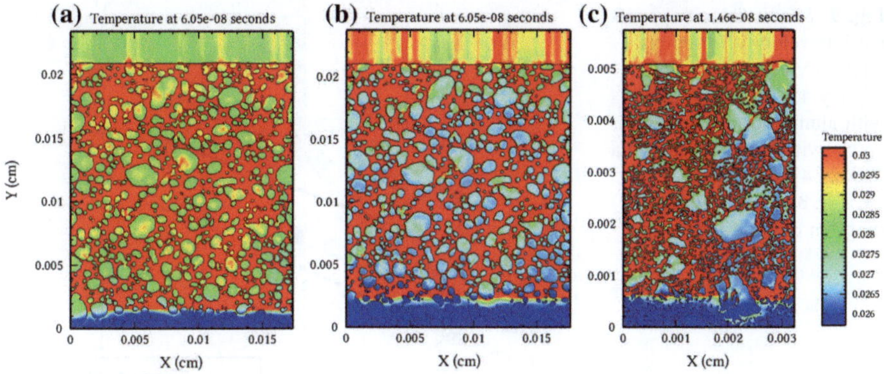

Fig. 8 The temperature present in each composite phase after an impact at 622 m/s. The composites are **a** aluminum-epoxy from micrograph **b** alumina-epoxy replacing the properties of the aluminum with alumina from (**a**), and **c** alumina-epoxy with a micrograph from [44]

be much faster than the initial shock wave [49, 50], which has been attributed to contact stresses between the alumina particles during loading and unloading [24], and a strong function of the particle velocity [49, 50] and volume fraction of particulate [44].

Similar composites containing aluminum and epoxy have been compared numerically to the alumina epoxy composites, as shown in Fig. 8. The alumina-epoxy composite shows increased temperature in the aluminum particles compared to the alumina particles, with it comparable to the epoxy binder, indicating that the aluminum particles have deformed during the shock event. Similar simulation of aluminum–tungsten-epoxy reveals that, again, the aluminum particles are deforming, where the tungsten particles are not [46]. However, the macroscopic response of these composites, as shown in Fig. 9, is extremely similar. For both the two component and three component systems, the slopes of the shock velocity-particle velocity curves are almost identical, indicating the dominance of the epoxy binder on the shock properties. The shift up or down of the curves is primarily due to the density differences in the composites.

7 Summary

Polymer-matrix particulate composites consist of individual particles of more than one material dispersed throughout and held together by a polymer binder. The mechanical and physical properties of the composite depend on the mechanical and physical properties of the individual components, particularly the polymer binder; their loading density; the shape and size of the particles; the interfacial adhesion; residual stresses; and matrix porosity. Systematic studies of the effects of volume fraction and microstructure on the behavior of these polymer-based

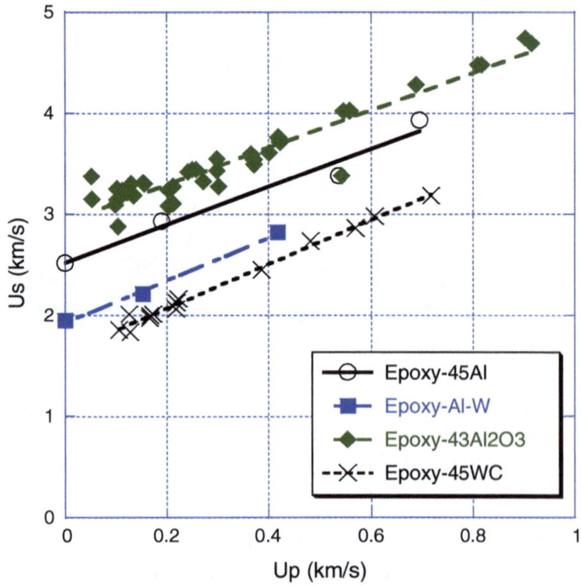

Fig. 9 Shock velocity versus particle velocity for epoxy with 45 vol.% aluminum (Epoxy-45 Al) [46], epoxy with aluminum and tungsten, epoxy with 43 vol.% alumina (Epoxy-43Al2O3) [20, 23, 24, 44], and epoxy with tungsten carbide [18]

composites are critical. The behavior of polymer-matrix particulate composites at intermediate to high strain rates has not been investigated in detail in the literature. The testing strain rate can greatly affect the behavior of these composites due to the dependency on rate dependant phase changes in the polymer binder. The intermediate strain rate behavior ($\sim 10^3$–10^4/s) is studied using a split Hopkinson pressure bar and Taylor impact experiments. Shock, or high strain rate, properties of these composite materials have been investigated using gas gun and explosive loading techniques. In this review, the methods for dynamic strength and shock properties in particulate composites were discussed. The literature for the high strain-rate deformation of particulate composites, primarily epoxy-based, were reviewed. Several key features and analyses for these composites were discussed, namely particle deformation and mechanical percolation thresholds. Finally, the available literature on the shock response of epoxy-based composites was discussed and summarized.

Acknowledgments The authors would like to thank both AFOSR and AFRL/RW for funding the work presented in this chapter and references. JLJ was employed by the Air Force Research Laboratory, Munitions Directorate and EBH was employed by the Georgia Institute of Technology when this work was performed and they gratefully acknowledge their support. Additionally, JLJ would like to thank her collaborators over the years—Mel Baer, John Borg, Eric Brown, Dana Dattelbaum, Richard Dick, Louis Ferranti, Andrew Fraser, Jason Foley, Wayne Richards, Stephen Sheffield, Clive Siviour, Jonathan Spowart, Gerrit Sutherland, Naresh Thadhani, Brad White. Opinions, interpretations, conclusions and recommendations are those of the authors and are not necessarily endorsed by the United States Air Force.

References

1. Eshelby, J.D.: The determination of the elastic field of an ellipsoidal inclusion, and related problems. Proc. R. Soc. Lond. A **241**(1226), 376–396 (1957)
2. Hashin, Z., Shtrikman, S.: A variational approach to the theory of the elastic behaviour of multiphase materials. J. Mech. Phys. Solids **11**(2), 127–140 (1963)
3. Mori, T., Tanaka, K.: Average stress in matrix and average elastic energy of materials with misfitting inclusions. Acta Metall. **21**(5), 571–574 (1973)
4. Ahmed, S., Jones, F.R.: A review of particulate reinforcement theories for polymer composites. J. Mater. Sci. **25**(12), 4933–4942 (1990)
5. Gray, G.T.: Classic Split Hopkinson Pressure Bar Testing. In: ASM Handbook Volume 8: Mechanical Testing and Evaluation, pp. 462–76. ASM International, Materials Park, OH (2000)
6. Jordan, J.L., Siviour, C.R., Foley, J.R., Brown, E.N.: Compressive properties of extruded polytetrafluoroethylene. Polymer **48**(14), 4184–4195 (2007)
7. Song, B., Chen, W., Montgomery, S.T., Forrestal, M.J.: Mechanical response of an alumina-filled epoxy at various strain rates. J. Compos. Mater. **43**(14), 1519–1536 (2009)
8. Taylor, G.I.: The use of flat-ended projectiles for determining dynamic yield stress. I. Theoretical considerations. Proc. R. Soc. Lond. A **194**(1038), 289–299 (1948)
9. Hutchings, I.M.: Estimation of yield stress in polymers at high strain-rates using GI Taylor's impact technique. J. Mech. Phys. Solids **26**(5-6), 289–301 (1978)
10. Rae, P.J., Brown, E.N., Clements, B.E., Dattelbaum, D.M.: Pressure-induced phase change in poly(tetrafluoroethylene) at modest impact velocities. J. Appl. Phys. **98**, 063521 (2005)
11. Sarva, S., Mulliken, A.D., Boyce, M.C.: Mechanics of Taylor impact testing of polycarbonate. Int. J. Solids Struct. **44**, 2381–2400 (2007)
12. Ferranti, L., Thadhani, N.N.: Dynamic mechanical behavior characterization of epoxy-cast Al^+ Fe2O3 thermite mixture composites. Meta. Mater. Trans. A **38A**, 2697–2715 (2007)
13. Martin, M., Hanagud, S., Thadhani, N.N.: Mechanical behavior of nickel + aluminum powder-reinforced epoxy composites. Mater. Sci. Eng. A **443**, 209–218 (2007)
14. White, B. W., Thadhani, N.N., Jordan, J.L., Spowart, J.E.: The effect of particle reinforcement on the dynamic deformation of epoxy-matrix composites. In: American Institute of Physics Conference Proceedings, vol. 1195, pp. 1245–1248 (2009)
15. Meyers, M.A.: Dynamic Behavior of Materials. Wiley, New York (1994)
16. Field, J.E., Walley, S.M., Proud, W.G., Goldrein, H.T., Siviour, C.R.: Review of experimental techniques for high rate deformation and shock studies. Int. J. Impact Eng. **30**, 725–775 (2004)
17. Jordan, J.L., Dattelbaum, D.M., Sutherland, G., Richards, D.W., Sheffield, S.A., Dick, R.D.: Shock equation of state of a multi-phase epoxy-based composite (Al–MnO^{2-} epoxy). J. Appl. Phys. **107**, 103528 (2010)
18. Vogler, T.J., Alexander, C.S., Wise, J.L., Montgomery, S.T.: Dynamic behavior of tungsten carbide and alumina filled epoxy composites. J. Appl. Phys. **107**, 043520 (2010)
19. Jordan, J.L., Ferranti, L., Austin, R.A., Dick, R.D., Foley, J.R., Thadhani, N.N., McDowell, D.L.: Equation of state of aluminum–iron oxide-epoxy composite. J. Appl. Phys. **10**(9), 093520 (2007)
20. Millett, J.C.F., Bourne, N.K., Deas, D.: The equation of state of two alumina-filled epoxy resins. J. Phys. D Appl. Phys. **38**, 930–934 (2005)
21. Sheffield, S.A., Gustavsen, R.L., Alcon, R.R.: In situ magnetic gauging technique used at LANL-method and shock information obtained. In: Shock Compression of Condensed Matter-1999, pp. 1043–1048. AIP Publishing, College Park, MD (2000)
22. Munson, D.E., May, R.P.: Dynamically determined high-pressure compressibilities of three epoxy resin systems. J. Appl. Phys. **43**(3), 962–971 (1972)

23. Anderson, M.U., Setchell, R.E., Cox, D.E.: Shock and release behavior of filled and unfilled epoxies. In: Furnish, M.D., Chhabildas, L.C., Hixson, R.S. (eds.) Shock Compression of Condensed Matter-1999, pp. 551–554 AIP Publishing, College Park, MD (2000)
24. Setchell, R.E., Anderson, M.U.: Shock-compression response of an alumina-filled epoxy. J. Appl. Phys. **97**, 083518 (2005)
25. Siviour, C.R., Walley, S.M., Proud, W.G., Field, J.E.: The high strain rate compressive behaviour of polycarbonate and polyvinylidene diflouride. Polymer **46**, 12546–12555 (2005)
26. Jordan, J.L., Foley, J.R., Siviour, C.R.: Mechanical properties of Epon 826/DEA epoxy. Mech. Time Dependant Mater. **12**, 249–272 (2008)
27. Clements, B.E., Mas, E.M.: Dynamic mechanical behavior of filled polymers. I. Theoretical developments. J. Appl. Phys. **90**(11), 5522–5534 (2001)
28. Clements, B.E., Mas, E.M.: Dynamic mechanical behavior of filled polymers. II. Applications. J. Appl. Phys. **90**(11), 5535–5541 (2001)
29. Jordan, J.L., Spowart, J.E., Richards, D.W.: Constitutive characterization of multi-constituent particulate composites. In: Proceedings of the Society for Experimental Mechanics Annual Meeting 2010. Springer, Heidelberg (2010)
30. Owens, A.T., Tippur, H.V.: A tensile split Hopkinson bar for testing particulate polymer composites under elevated rates of loading. Exp. Mech. **49**, 799–811 (2009)
31. Predecki, P., Barrett, C.: Stress measurement in graphite/epoxy composites by X-Ray diffraction from fsillers. J. Compos. Mater. **13**, 61–71 (1979)
32. Benedikt, B., Lewis, M., Rangaswamy, P.: An analysis of internal strains in unidirectional and chopped graphite fibre composites based on x-ray diffraction and micro Raman spectroscopy measurements. In: International Conference on Computational Methods and Experiments in Materials Characterization, pp. 13–22. Elsevier, Amsterdam (2005)
33. Benedikt, B., Lewis, M., Rangaswamy, P.: Measurement and modeling of internal stresses at microscopic and mesoscopic levels using micro-Raman spectroscopy and X-ray diffraction. Powder Diffr. **21**(2), 118–121 (2006)
34. Benedikt, B., Lewis, M., Rangaswamy, P., Kumosa, M., Predecki, P., Kumosa, L., Gentz, M.: Residual stress analysis in aged graphite/PMR-15 composites using X-ray diffraction. Mater. Sci. Eng. A **421**, 1–8 (2006)
35. Quane, S.L., Russell, J.K.: Bulk and particle strain analysis in high-temperature deformation experiments. J. Volcanol. Geoth. Res. **154**, 63–73 (2006)
36. Segurado, J., Llorca, J.: Computational micromechanics of composites: The effect of particle spatial distribution. Mech. Mater. **38**, 873–883 (2006)
37. White, B.W., Spowart, J.E., Jordan, J.L., Thadhani, N.N.: Strain rate effects on the deformation behavior of particles in epoxy-based composites. Presentation at 2010 TMS Annual Meeting & Exhibition (2010)
38. Kirkpatrick, S.: Percolation and conduction. Rev. Mod. Phys. **45**, 574–588 (1973)
39. Lange, F.F., Atteraas, L., Zok, F., Porter, J.R.: Deformation consolidation of metal powders containing steel inclusions. Acta metallurgica et metallica **39**(2), 209–219 (1991)
40. Lekatou, A., Faidi, S.E., Lyon, S.B., Newman, R.C.: Elasticity and fracture in particulate composites with strong and degraded interfaces. J. Mater. Res. **11**(5), 1293–1304 (1996)
41. Novak, I., Krupa, I., Chodak, I.: Relation between electrical and mechanical properties in polyurethane/carbon black adhesives. J. Mater. Sci. Lett. **21**, 1039–1041 (2002)
42. Chodak, I., Krupa, I.: 'Percolation effect' and mechanical behavior of carbon black filled polyethylene. J. Mater. Sci. Lett. **18**, 1457–1459 (1999)
43. Verbeek, C.J.R.: Effect of percolation on the mechanical properties of sand-filled polyethylene composites. J. Thermoplast. Compos. Mater. **20**, 137–149 (2007)
44. Setchell, R.E., Anderson, M.U., Montgomery, S.T.: Compositional effects on the shock-compression response of alumina-filled epoxy. J. Appl. Phys. **101**, 083527 (2007)
45. Ferranti, L.: Ph.D dissertation, Georgia Institute of Technology (2007)
46. Jordan, J.L., Herbold, E.B., Sutherland, G., Fraser, A., Borg, J., Richards, D.W.: Shock equation of state of multi-constituent epoxy-metal particulate composites. J. Appl. Phys. (2011 in press)

47. Carter, W.J., Marsh, S.P.: Hugoniot equation of state of polymers. LA-13006-MS, Los Alamos National Laboratory (1999)
48. Millett, J.C.F., Bourne, N.K., Barnes, N.R.: The behavior of an epoxy resin under one-dimensional shock loading. J. Appl. Phys. **92**, 6590 (2002)
49. Anderson, M.U., Setchell, R.E., Cox, D.E.: Effects of initial temperature on the shock and release behavior of filled and unfilled epoxies. In: Furnish, M.D., Thadhani, N.N., Horie, Y. (eds.) Shock Compression of Condensed Matter-2001, pp. 669–672 (2002)
50. Munson, D.E., Boade, R.R., Schuler, K.W.: Stress-wave propagation in Al_2O_3—epoxy mixtures. J. Appl. Phys. **49**, 4797–4807 (1978)
51. Chhabildas, L.C., Swegle, J.W.: On the dynamical response of particulate-loaded materials. I. Pressure-shear loading of alumina particles in an epoxy matrix. J. Appl. Phys. **53**(2), 954–956 (1982)
52. Drumheller, D.S.: On the dynamical response of particulate-loaded materials. II. A theory with application to alumina particles in an epoxy matrix. J. Appl. Phys. **53**(2), 957–969 (1982)
53. Millett, J.C.F., Bourne, N.K., Deas, D., Montgomery, S.T.: The deviatoric response of an alumina filled epoxy composite during shock loading. J. Appl. Phys. **102**, 063518 (2007)

FEM Modelling of the Continuous Combined Drawing and Rolling Process for Severe Plastic Deformation of Metallic Materials

C. J. Luis, D. Salcedo, R. Luri, J. León and I. Puertas

Abstract In the last years, new severe plastic deformation (SPD) processes have been developed in order to produce very high values of deformation in the materials to be processed, with only small changes in the size of parts. This is not possible through conventional thermo-mechanical processes, in which the increase up to the above-mentioned values of deformation is usually associated with a change in their geometry. As is well-known, with enough accumulation of plastic deformation, a new submicrometric or even nanometric grain structure substitutes the former. Given that the grain size in metallic materials has a great deal of influence on their mechanical properties, the refinement of this grain size provides enormous technological advantages. For instance, at low values of temperature, a fine grain size can increase mechanical strength, hardness, fracture toughness and the material fatigue limit. Furthermore, at high values of temperature, the alloys with an ultrafine grain size may exhibit a superplastic behaviour and thus, the ability to undergo very high values of deformation with no damage to the material. The continuous combined drawing process in angular channels (CCDR) is a new concept of severe plastic deformation process (SPD), developed by researchers belonging to the Public University of Navarre and based on the patent (ES 2224787). In the present work, finite element analysis will be employed in order to study not only the strain

C. J. Luis (✉) · D. Salcedo · R. Luri · J. León · I. Puertas
Mechanical, Energetics and Materials Engineering Department, Public University of Navarre, Campus de Arrosadia s/n 31006 Pamplona, Spain
e-mail: cluis.perez@unavarra.es

D. Salcedo
e-mail: daniel.salcedo@unavarra.es

R. Luri
e-mail: rodrigo.luri@unavarrra.es

J. León
e-mail: javier.leon@unavarra.es

I. Puertas
e-mail: inaki.puerta@unavarra.es

N. Bonora and E. Brown (eds.), *Numerical Modeling of Materials Under Extreme Conditions*, Advanced Structured Materials 35, DOI: 10.1007/978-3-642-54258-9_2,

distribution in the processed materials but also the homogeneity of the introduced strain. Moreover, experimental results will be compared to that obtained by using FEM. With the present work, it will be shown that it is feasible to achieve a process with possible industrial application, making the continuous processing of metallic materials in angular channels by SPD possible.

Keywords SPD · CCDR · FEM

1 Introduction

In this chapter, the results obtained by the finite element method of the so-called continuous combined drawing and rolling (CCDR) are shown by means of using several process conditions and types of materials. This process is a new concept of severe plastic deformation processes (SPD), developed by researchers belonging to the Public University of Navarre and based on the patent (ES 2224787) [1].

As can be observed in the state-of-the-art works recently published, a growing tendency exists in these last years towards the development of processes which make it possible to introduce very high values of plastic deformation within the materials to be processed with no significant change of their cross-section. This is not possible through conventional thermo-mechanical processes, in which the increase until the above-mentioned values of deformation is usually associated with a change in their geometry [2–5].

The aim of introducing very high values of plastic deformation within the processed materials is to achieve the appearance of a new submicrometric or even nanometric grain structure which will gradually substitute that existing initially, where this fact is favoured by an increase in the temperature of the process. Moreover, the fact of achieving a submicrometric structure presents several technological advantages, as this will have a direct influence on the mechanical properties of the new so-processed materials. For instance, at low values of temperature, a fine grain size can increase mechanical strength, hardness, fracture toughness and the material fatigue limit. Furthermore, at high values of temperature, the alloys with a ultrafine grain size may exhibit a superplastic behaviour and thus, the ability to undergo very high values of deformation with no damage for the material [6, 7].

Nevertheless, few processes which are able to take advantage of the benefits of the plastic deformation in angular channel exist, which could indicate the technological difficulty associated with the development of a process capable of operating continuously in order to impart high values of plastic deformation to metallic materials by means of an angular channel.

Among the discontinuous severe plastic deformation (SPD) techniques that have been developed in the last years to manufacture massive materials with submicrometric or even nanometric grain size, the first and more widely known is the Equal Channel Angular Extrusion or Pressing (ECAE or ECAP), which was

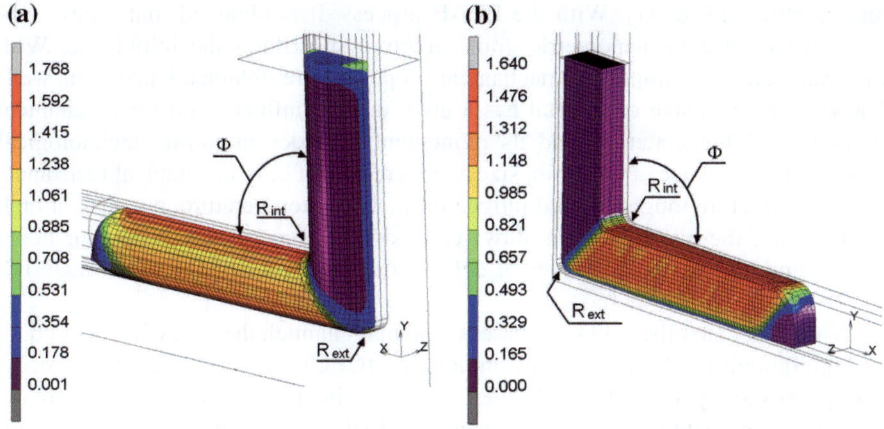

Fig. 1 Total equivalent plastic strain in the ECAE process **a** circular die, **b** square die

developed by Segal and his collaborators in the former Soviet Union, at the Institute for Metals Superplasticity Problems [8].

The angular channel extrusion consists in the compression of the billet of a material through a die with two channels which intersect at an angle varying generally between 90° and 135°, where the billet is extruded through the exit channel and this is the same size as the entrance channel. Figure 1 shows a finite element simulation of the ECAE process considering circular and square cross-section. When the material crosses the intersection between the two channels of the die, this undergoes a very intense shear plastic deformation ($\varepsilon > 1$) in a single passage. The material can be deformed with no fracture up to ($\varepsilon > 1$) due to the high value of hydrostatic pressure which exists inside the die channels when the extrusion process is carried out [9].

Although it is possible to achieve deformation values similar to those obtained in ECAE ($\varepsilon > 1$) by means of other conventional plastic deformation processes, such as drawing or rolling, the final values of section or thickness are reduced significantly, as such deformation increase is associated with a geometrical change. If it is taken into consideration that the ECAE process can be repeated several times as the part dimensions do not change, it is possible to accumulate a higher value if deformation, that is ($\varepsilon \gg 1$), which is translated into the fact that if it is intended to reach by conventional thermomechanical processes similar deformation values, the final dimensions would be so small that this would limit its subsequent application as structural element, not being able even to perform it [10]. As a consequence of this, severe plastic deformation processes have a great deal of interest, as the accumulation of deformation inside the processed material is achieved with no significant change in its cross-section.

The ECAE process allows us to manufacture highly deformed materials with ultrafine grain size and free of residual porosity. The process has been applied to a large number of materials such as copper, nickel, aluminium alloys and steel, among others [3, 8–37]. With the ECAE process, it is obtained that a new submicrometric or even nanometric grain structure substitutes the initial one. With this, materials with improved mechanical properties are obtained since, as is well-known, the grain size of a metal has a great deal of influence on the mechanical properties of the materials and its refinement provides important technological benefits. For instance, a fine grain size can increase not only mechanical resistance, hardness, fracture toughness and fatigue limit, at low temperature, but also, at high temperature, the ultrafine grain alloys can exhibit a superplastic behaviour being able to undergo very high values of deformation with no damage to the material [6, 7].

With respect to other SPD processes in angular channel, the following ones could be highlighted: Cyclic Extrusion Compression (CEC) [38, 39], where this process was proposed by Korbel and Richert [38], Cyclic Extrusion and Compression (CEC), denominated CECRE (C shape Equal Channel Reciprocating Extrusion) [40] and Twist Extrusion, developed by Akbari et al. [41], among others.

Regarding other SPD processes which do not use an angular channel, the following ones stand out: the combined process of high pressure and torsion (High Pressure Torsion or HPT) [18, 42–46], Repetitive Corrugation and Straightening (RCS) [47, 48] and the process denominated as Multipass Coin Forging (MCF), which is similar to RCS [49], except for the arrangement and the geometry of the dies.

In spite of the great interest that discontinuous SPD processes present, as they allow us to obtain nanostructured materials, their industrial applications have not been as remarkable as could be expected, being in this aspect the ECAE process the one that presents a higher number of applications, such as those carried out by Ferrase et al. [19], Valiev et al. [50], Zhu et al. [51], Latysh et al. [52] and Tanaka et al. [53], among others.

In relation to the rest of the above-mentioned discontinuous SPD processes, such as HPT or CGP, the fact that they are discontinuous and relatively low and that they present a high value of heterogeneity in the deformation along with the fact that the size of the processed parts is not high enough could be what has limited their industrial applications.

Although the ECAE process presents diverse applications, the length of the billets is limited due to the buckling the punch can undergo if values of length more than ten times the value of the material diameter are employed. Furthermore, the process velocity is low since the values for the strain rate are not very high and it is necessary to open and close the die each time a new extrusion is needed to be performed.

Although several industrial applications have been carried out with this process, its importance is not as great as could be expected taking into consideration the significant improvement the so-processed materials imply.

In the present work, finite element analysis will be employed in order to study not only the strain distribution in the processed materials but also the homogeneity of the introduced strain in the CCDR process [1]. With the present work, it will be shown that it is feasible to achieve a process with possible industrial application, making possible the continuous processing of metallic materials in angular channel by SPD.

2 Description of the CCDR Process

In the CCDR process (Continuous Combined Drawing and Rolling), based on the patent ES-2002-01163 [1], a pushing force is generated at the entrance of the die by the effect of two rolling mills, whereby this is combined with a drawing force. By means of this process, it is possible to manufacture continuous materials through severe plastic deformation in angular channels [1, 54, 55]. In these days, a prototype has been developed in the Public University of Navarre, based on the previous patent for the continuous processing of sheet.

Figures 2 and 3 show different configurations for the implementation of the patented process [1], depending on the requirement to manufacture sheet or wire and where the rolling mills are not drawn to scale.

As can be observed, a compression force is generated at the entrance of the die by the employment of rolling mills and, simultaneously to the rolling process, a drawing process exists by which a force drawing the material is developed at the exit of the angular channel die. As a consequence of this, the stress values required for the rolling process are reduced as the device at the exit will help the material to pass through the angular channel die. Moreover, between passage and passage, the sheet can be rotated in order to homogenise the deformation pattern which is obtained along the cross-section of the processed material.

Among the different possible CCDR set-ups, some of them are shown in the following figures. Sheet and cylindrical bars or with other types of cross-section can be processed. Depending on the type of material to be processed, diverse applications can be found. The most outstanding ones are as follows: rolled structural shapes, wires, shapes for diverse applications, plane sheet and substituting steel parts for parts made of high resistance aluminium with mechanical properties improved by the refinement of grain that will be mentioned below.

Figure 4 shows the development of the continuous process of plastic deformation by angular channel for manufacturing sheet, as was shown in the operation sketch from Fig. 2. As can be observed, it is a plastic deformation process by angular channel which operates continuously, avoiding the buckling problem associated with the ECAE process, which limited the size of the processed parts. Figure 5 shows the material obtained with the process that has been developed.

The process denominated as "Continuous processing of metallic materials by means of plastic deformation with polyangular channel", based on the patent Pat. 2224787 [1] or, in English, Continuous Combined Drawing and Rolling (CCDR) was proposed by researchers from the Public University of Navarre in collaboration with the "Centro de Estudios e Investigaciones Tecnológicas" from Guipúzua. The patented process presents the novelty of combining an extrusion force with a pushing system, composed of rolling mills. By the combined action of both processes, it is possible to process material continuously through a die which presents a configuration with angular channel.

The novelty of the CCDR process [1] in comparison to other continuous developed SPD processes consists in combining a pushing system in addition to a

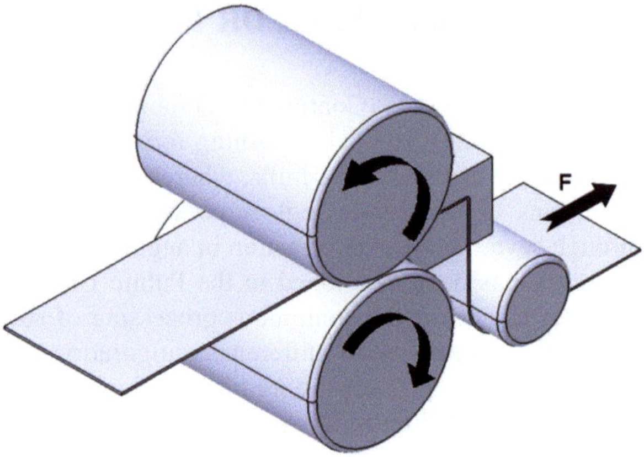

Fig. 2 Operation sketch of the CCDR process for manufacturing sheet [1]

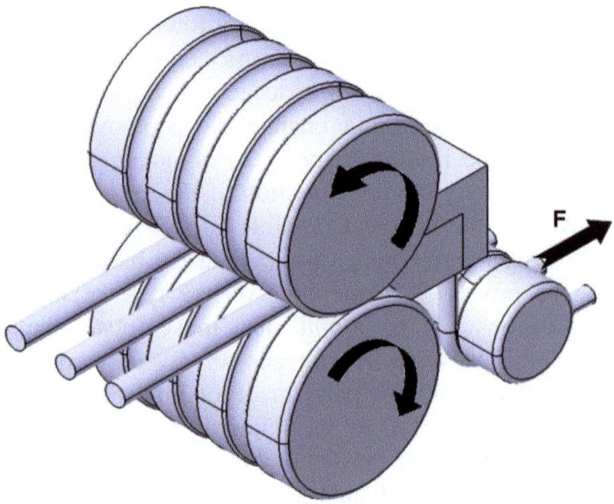

Fig. 3 Operation sketch of the CCDR process for manufacturing wire [1]

compression system that thus enables the material flow in the angular channel die, preventing the material from wrinkling at the entrance. Figures 2 and 3 show the fundamentals of the Continuous Combined Drawing and Rolling (CCDR). As can be seen in these figures, a compression force is generated at the entrance of the die by employing rolling mills and, simultaneously to the rolling process, a drawing process exists by which a force drawing the material is developed at the exit of the angular channel die. As a consequence of this, the stress values required for the rolling process are reduced significantly as the device at the exit will help the material to pass through the angular channel die. In this way, the wrinkling of the material is avoided at the entrance of the ECAE dies. With this process, it is possible to continuously manufacture materials through severe plastic deformation with an angular channel.

Figure 5 shows the material obtained with the process that has been developed. It is important to ensure that the material velocity at the exit is the same as the material velocity at the entrance of the die channel and, for this purpose, a series of

Fig. 4 Equipment developed for implementing a continuous process of plastic deformation by angular channel for manufacturing sheet, based on the patent proposed by Luis et al. (Pat. 2224787, 2002) [1]. The prototype is installed in the facilities of the Department of Mechanical, Energetics and Materials Engineering from the Public University of Navarre

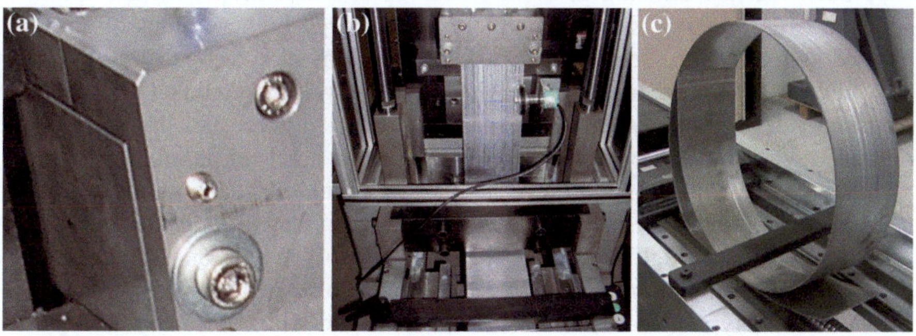

Fig. 5 View in detail of **a** the angular channel die, **b** measurement of the material velocity at the exit of the die, and **c** material processed by the CCDR equipment based on [1]

encoders are placed in order to determine the real velocity of the sheet. In this way, it is possible to regulate the CCDR process, not allowing the sheet to be stretched or wrinkled at the entrance of the die channel.

This equipment combines a rolling process with a drawing process, as is pointed out in Patent 2224787 [1]. In this way, the rolling mills are responsible for the force necessary to compress the material until the intersection of the die angular channel, whereby the gripping system has the aim of drawing the material at the exit of the die, in such a way that the material velocity at the entrance of the angular channel coincides with the material velocity at the exit of the latter. As a result, no significant necking of the processed material is produced.

In order to be able to control the sheet velocity at different points during the process, a series of measurement sensors have been installed. With the performance of these measurements, it is possible to modify via PC both the rolling or the drawing velocity, as convenient and in order to prevent not only a possible bulging of the material between the rolling mills and the die but also a drastic reduction of the material cross-section at the exit.

3 Finite Element Modelling

In the present section, some simulations performed on the CCDR process, presented in the previous section, are shown. In order to perform them, an aluminium alloy of type AA8011 has been selected. With the aim of analysing the processing of the material by CCDR through the use of finite element method (FEM), the process conditions that are shown in Table 1 have been selected. The sheet width was considered to be 130 mm and it was maintained constant in all the simulations.

With respect to the elements, their size is of 0.05×0.25 mm for the simulations with a single sheet and of 0.05×0.5 mm for those with two sheets. The dies have been meshed utilising division of curves and automesh with the option of advance front, making the zones which will undergo higher values of stress be meshed as fine as the sheet element so that situations of nodes penetration do not take place (0.05 mm). In the same way, it is very important that the nodes in contact are 10 or more so that the peaks in the values of force and stress, which are produced when the sheet is being processed, are as small as possible.

In Fig. 6, an image of the die mesh can be observed. Plane strain conditions have been employed as the sheet width is much higher than its thickness. This approach is good enough and it saves calculation time, in relation to that necessary to perform the simulation considering a tridimensional geometry.

An element type Quad 4, which is a quadrilateral with four integration points, has been utilised. The number of elements is 54,000. As was previously-mentioned, plane strain conditions have been assumed since the sheet width is much higher that the sheet thickness.

With respect to the material employed in the FEM simulations, the aluminium alloy AA8011, whose flow stress is depicted in Fig. 7, has been selected.

In order to estimate the damage the sheet undergoes during the process, a model type Cockroft–Latham [56] has been taken into account. The damage threshold,

Table 1 Simulations performed

Sheet of 2.6 mm, rolling 2 mm, die channel 2.2 mm, internal radius 2 mm
Sheet of 2.6 mm, rolling 2 mm, die channel 2.2 mm, internal radius 1.5 mm
Sheet of 5.2 mm, rolling 4 mm, die channel 4.2 mm, internal radius 2 mm
Sheet of 5.2 mm, rolling 4 mm, die channel 4.2 mm, internal radius 1.5 mm
Sheet of 7.8 mm, rolling 6 mm, die channel 6.2 mm, internal radius 2 mm
Sheet of 7.8 mm, rolling 6 mm, die channel 6.2 mm, internal radius 1.5 mm
Two sheets, each one of 2.6 mm, rolling 4 mm, die channel 4.2 mm, internal radius 2 mm
Two sheets, each one of 2.6 mm, rolling 4 mm, die channel 4.2 mm, internal radius 1.5 mm
Three sheets, each one of 2.6 mm, rolling 6 mm, die channel 6.2 mm, internal radius 2 mm
Three sheets, each one of 2.6 mm, rolling 6 mm, die channel 6.2 mm, internal radius 1.5 mm

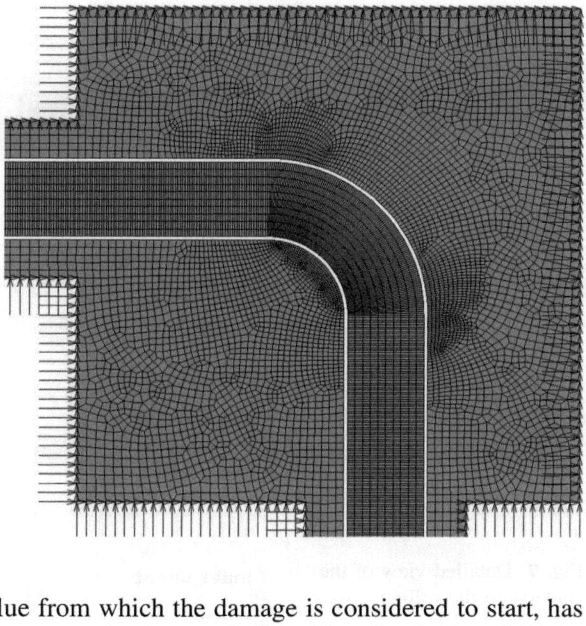

Fig. 6 Meshing of both angular channel die and processed material

which is the deformation value from which the damage is considered to start, has been selected as zero, that is to say, the damage is considered to begin when the material is deformed.

In relation to the contacts employed, except for the case of the sheets and the dies, which are all deformable bodies, the rest of the bodies used in the simulations are considered to be rigid. Figures 8 and 9 show the contacts that are taken into account in the performed simulations.

The velocity of the process is 20 mm/min and the rolls rotate at 0.004 rad/s. It is important to point out that the drawing velocity is slightly higher than the velocity of the rolls in order to avoid the buckling of the material at the entrance of the angular channel dies.

Regarding the contacts, a distance tolerance of 0.0025 mm is considered to detect the contact situation and a bias factor of 0.9 for the sensibility of a rigid

Fig. 7 AA8011 flow stress

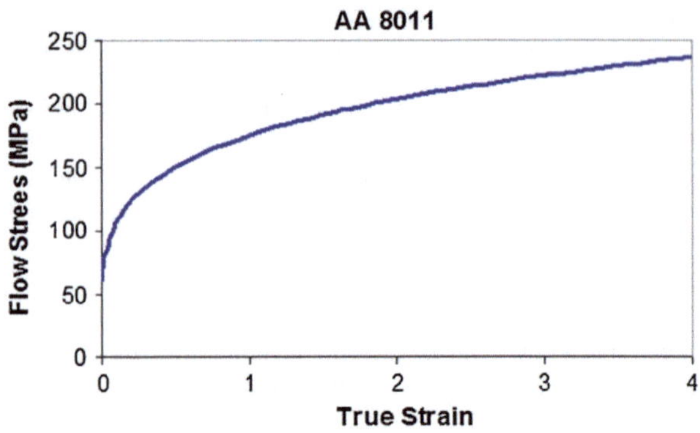

Fig. 8 Contacts die-material

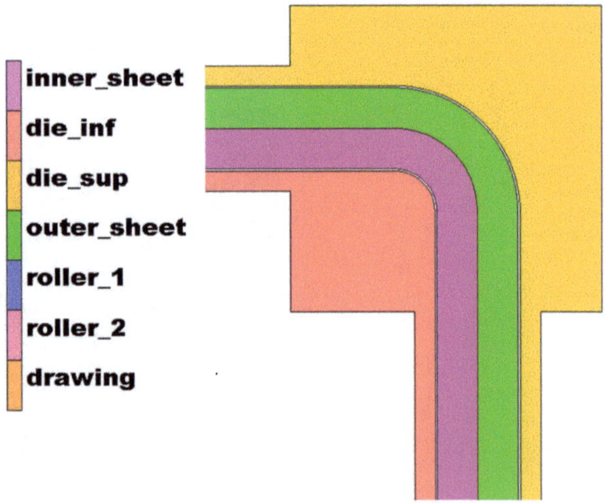

Fig. 9 Detailed view of the contacts in the rolling

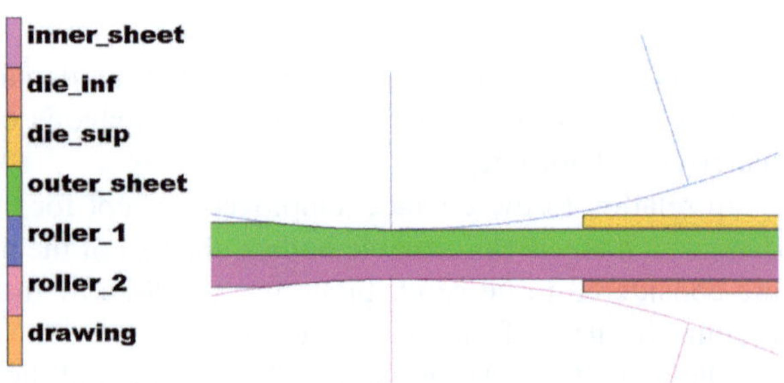

body to be crossed and of 0 when there is a contact between deformable bodies (0 means that when the contact is detected, the interval fixed by the distance tolerance is symmetric, which this is very useful when the bodies in contact are

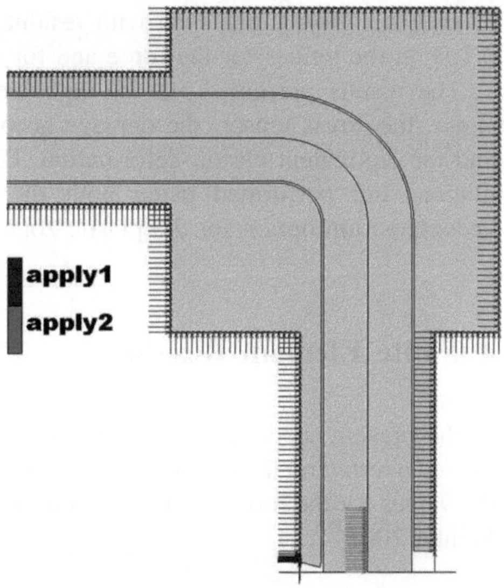

Fig. 10 Detailed view of the angular channel die

both deformable). Moreover, the initial stress in the contacts is not taken into account in order to avoid subsequent problems.

The friction model employed is that of type Shear [57], with a velocity of relative displacement of 0.025 mm/s. The values for the friction coefficients are as follows: 0.1 between sheets; 0.9 between the rolls and the sheet, and 0.1 between the dies and the sheet. The body which draws from the sheets is glued to an end of the latter.

The detection method for the contact between deformable bodies (useful in the case of two sheets and in the case of one sheet against any of the dies) is double sided, that is, it checks the contact at the nodes of the two bodies, which slows the simulation but is essential for the good running of this. The criterion of separating the nodes from the lines which limit the rigid bodies is considered through the relative stress, where if this changes the 0.1 %, the nodes are separated from the rigid body.

When the dies model is changed from a static rigid body to a deformable body, it is necessary to introduce restrictions in the nodes in order to avoid their displacements. Restriction of movement in axes "x" and "y" is applied to the external nodes of the dies. As well as this boundary condition, in the simulations performed with more than one sheet, a restriction of movement in axis "x" is applied to the first nodes in contact with the rigid body, which model the tension force exerted over the sheet at the exit of the angular channel. Figure 10 shows the previously mentioned boundary conditions as "apply1" and "apply2", respectively.

The iterative process employed to linearise the problem is the full Newton–Raphson method [57] with 50 possible recycles and with the possibility to continue the calculation although non-positive definite matrices appear. In order to solve the equations in each step of the simulations, direct sparse is utilised. The convergence of each iteration is achieved with residuals or displacements with a tolerance of 0.1 % in the values for the force and for the relative displacement.

The results presented are the equivalent plastic deformation, the von Mises stress, the stress tensor, the damage according to the Cockroft–Latham's model and the equivalent elastic deformation. For high values of deformation, the simulations are performed using both the Lagrange's method and the additive decomposition option for the plastic zone.

4 Finite Element Results

In the present section, the main results obtained in relation to the plastic deformation exerted on the material, the accumulated damage, the stresses in the die and the values for the force are assessed employing the FEM simulation software Marc Mentat 2008™.

4.1 Total Equivalent Plastic Strain

The aim of the CCDR process is to impart to the processed material high values of deformation with no significant change of its cross-section, in such a way that it is possible to accumulate higher deformation values. In the present section, the results obtained regarding the accumulation of plastic deformation inside the processed sheet are shown.

As was above-mentioned, by CCDR, the cross-section is reduced slightly in order to prevent the sheet from being accumulated at the entrance of the die. The temperature is being applied to the process to ensure that the section does not vary as it is possible to work with similar velocities of pushing and drawing. In addition to this, it is possible to reduce the heterogeneity that appears when the material is processed twice using route C between passes, where this route consists in rotating 180° the processed sheet in such a way that the zone which is first in contact with the inner radius of the die is in the next passage in contact with the outer radius, thus increasing the deformation exerted on the material and its homogeneity. This has not been included in this chapter because it does not form part of its objectives but it will be shown in a future technical work.

The values for the plastic deformation accumulated in the material as a consequence of the rolling and of the deformation in the angular channel have been assessed. A comparison of the deformation in the same sheet has been made when the inner radius and the sheet thickness vary. Moreover, the deformation values when a single sheet or when various sheets with the same thickness of the latter pass through the same die. In the images shown in Fig. 11a–g, the deformation imparted to the material can be observed when modifying the process conditions.

Taking the obtained results into account, the deformation diminishes when the internal radius of the lower die (Rint) is increased.

In the case of Fig. 12a, corresponding to a sheet of 2.6 mm thickness rolled up to 2 mm, it can be observed that the deformation obtained with the lowest inner radius is always higher than that obtained with the highest inner radius. The maximum value is located at the zone closer to the inner radius, a decrease is observed in the central zone and a new increase at the other end of the sheet. For 2 mm, the maximum value is of 1.3 whereas in the case of the inner radius 1.5 mm, the deformation value is of 1.5, which means a 15 % higher. Figure 12b, which is the case of one sheet of 5.2 mm thickness rolled up to 4 mm, shows that the deformation has a pattern similar to the previous case. The difference between the deformation obtained with the different radii is lower. The maximum value is attained for the radius of 2 mm although it is the intermediate zone of the sheet where a higher value of deformation is observed when a lower inner radius is used. Furthermore, Fig. 12c shows the deformation of one sheet of 7.8 mm thickness rolled up to 6 mm and later on, processed by CCDR. As can be observed, when the internal radius of the angular channel die is decreased, the achieved value of deformation is higher.

In order to analyse if there exists a change in the deformation imparted when a higher number of sheets with the same thickness is processed simultaneously, the simulations previously shown were performed, whose comparison is depicted in Fig. 13. In the study case for the three sheets, the existence of hydrostatic pressure in the central sheet was considered in order to diminish the damage and to increase the deformation value. In the following graphs, the attained results are shown.

In Fig. 13a and b, it can be observed that the deformation with a single sheet is always higher than that in the case of two sheets with the equivalent section. A border effect is seen in the case of the sheets processed simultaneously but it is not enough to improve that obtained with a single sheet of processed material. The same behaviour can be observed in Fig. 13c and d. Both inner and outer sheets provide deformation values much lower than those attained when a single sheet is processed. On the contrary, the intermediate sheet has deformation values similar to those obtained with a single sheet, presenting the advantage of having a much higher homogeneity of plastic deformation. Nevertheless, it can be seen that the central sheet reduces its section more than both, the internal and the external ones, as can be observed in Fig. 11g.

The thickness of the processed sheet is important for the possible applications of the material processed by CCDR. Figure 14a and b show the deformation values attained for sheets with a starting thickness of 2.6, 5.2 and 7.8 mm.

As can be observed in Fig. 14a and b, when the thickness is increased, the homogeneity and the mean deformation in the section increase. The maximum value is achieved for the sheet with an intermediate value of thickness although it seems that the sheet with the highest thickness presents a decrease in the central zone of a lesser degree.

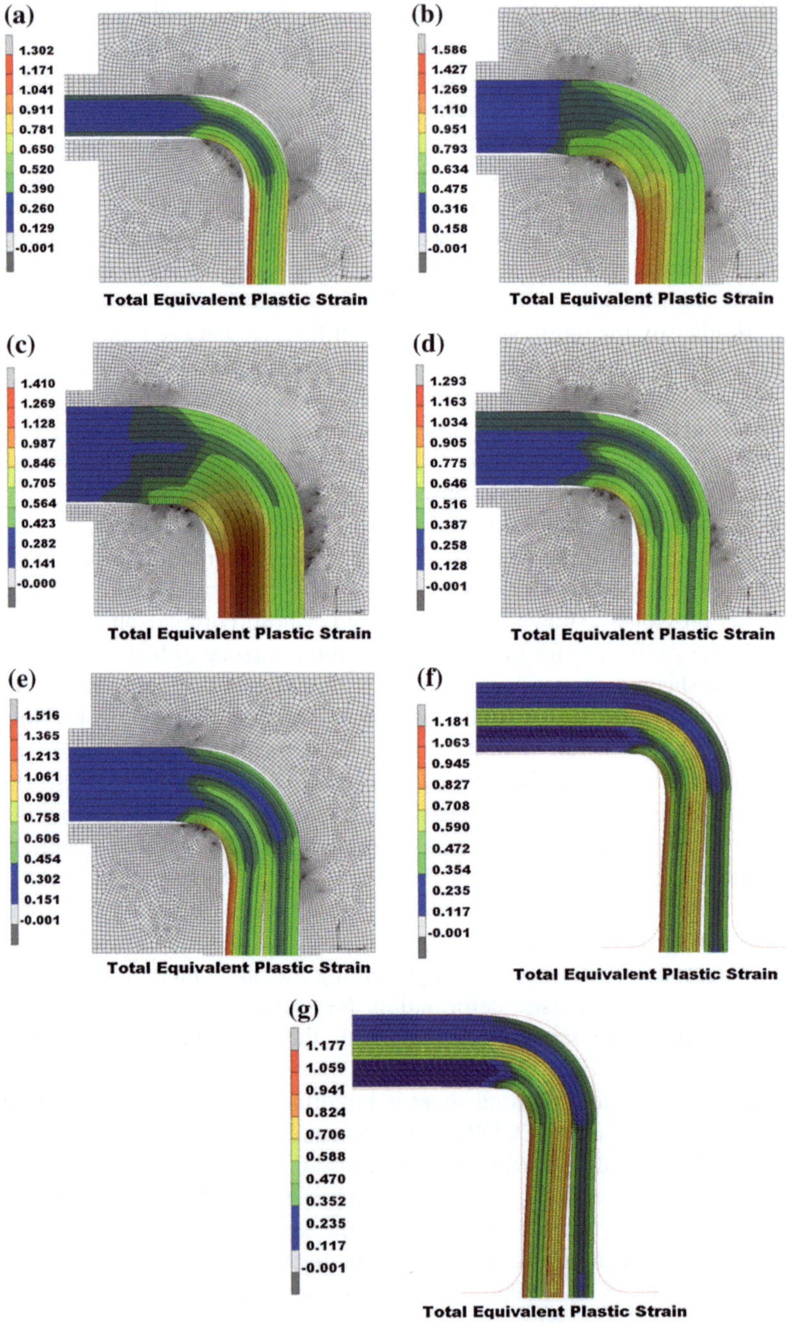
(a)
1.302
1.171
1.041
0.911
0.781
0.650
0.520
0.390
0.260
0.129
-0.001
Total Equivalent Plastic Strain
(b)
1.586
1.427
1.269
1.110
0.951
0.793
0.634
0.475
0.316
0.158
-0.001
Total Equivalent Plastic Strain
(c)
1.410
1.269
1.128
0.987
0.846
0.705
0.564
0.423
0.282
0.141
-0.000
Total Equivalent Plastic Strain
(d)
1.293
1.163
1.034
0.905
0.775
0.646
0.516
0.387
0.258
0.128
-0.001
Total Equivalent Plastic Strain
(e)
1.516
1.365
1.213
1.061
0.909
0.758
0.606
0.454
0.302
0.151
-0.001
Total Equivalent Plastic Strain
(f)
1.181
1.063
0.945
0.827
0.708
0.590
0.472
0.354
0.235
0.117
-0.001
Total Equivalent Plastic Strain
(g)
1.177
1.059
0.941
0.824
0.706
0.588
0.470
0.352
0.235
0.117
-0.001
Total Equivalent Plastic Strain

◄ **Fig. 11** Accumulated value of plastic deformation inside the material when varying the CCDR process conditions. **a** CCDR simulation (increment 1,500) in a die with an inner radius of 2 mm and one sheet with a thickness of 2 mm after rolling. **b** CCDR simulation (increment 1,500) in a die with an inner radius of 2 mm and one sheet with a thickness of 4 mm after rolling. **c** CCDR simulation (increment 1,500) in a die with an inner radius of 2 mm and one sheet with a thickness of 6 mm after rolling. **d** CCDR simulation (increment 1,500) in a die with an inner radius of 2 mm and two sheets with a thickness of 2 mm after rolling. **e** CCDR simulation (increment 1,500) in a die with an inner radius of 1.5 mm and two sheets with a thickness of 2 mm after rolling. **f** CCDR simulation (increment 1,650) in a die with an inner radius of 2 mm and three sheets with a thickness of 2 mm after rolling. **g** CCDR simulation (increment 1,650) in a die with an inner radius of 1.5 mm and three sheets with a thickness of 2 mm after rolling

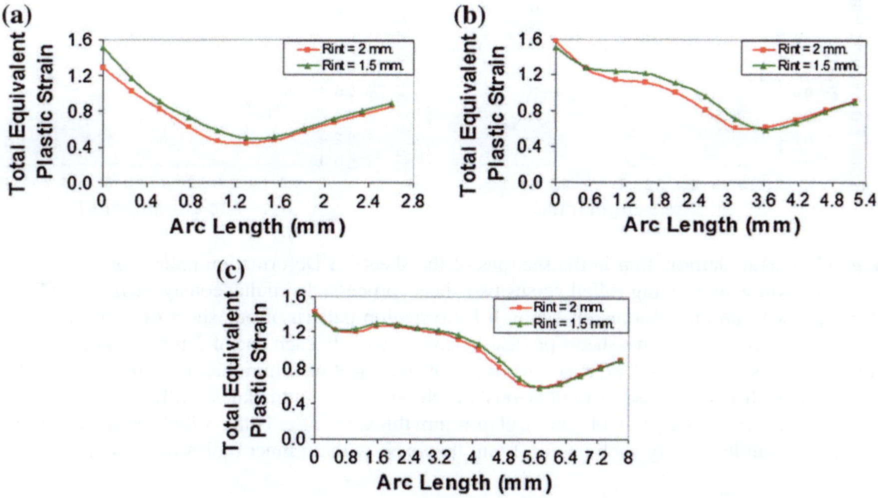

Fig. 12 Plastic deformation in a section of the sheet. **a** Deformation pattern of one sheet with a thickness of 2 mm after rolling. **b** Deformation pattern of one sheet with a thickness of 4 mm after rolling. **c** Deformation pattern of one sheet with a thickness of 6 mm after rolling

4.2 Analysis of the Damage Imparted to the Processed Material

Although with the CCDR process it is intended to impart very high values of plastic deformation to the processed materials, it is essential that the imparted damage is as low as possible in order to diminish the appearance of cracks limiting the subsequent applications of the processed material.

The damage is the progressive process by which the materials break subjected to a specific load. In a microscale level, this is the accumulation of microstresses in the neighbourhood of defects and the breaking of bonds, which generates micro-cracks that initiate one crack [58]. Damage is not an observable variable and a model definition is required in order to measure it [59]. Different approaches have

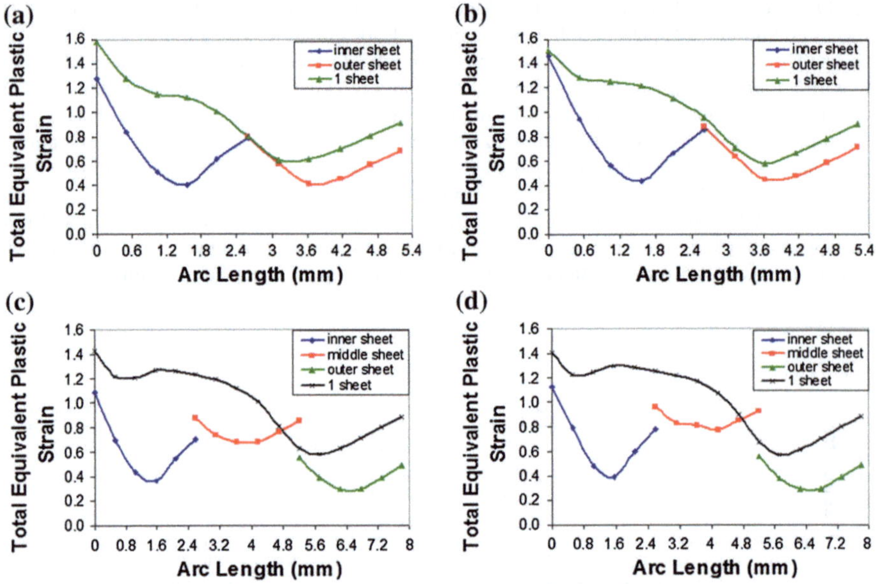

Fig. 13 Plastic deformation in the sections of the sheets. **a** Deformation pattern of one sheet of 4 mm thickness after being rolled versus two sheets processed simultaneously each one of 2 mm thickness with an inner radius of 2 mm. **b** Deformation pattern of one sheet of 4 mm thickness after being rolled versus two sheets processed simultaneously each one of 2 mm thickness with an inner radius of 1.5 mm. **c** Deformation pattern of one sheet of 6 mm thickness after being rolled versus three sheets processed simultaneously each one of 2 mm thickness with an inner radius of 2 mm. **d** Deformation pattern of one sheet of 6 mm thickness after being rolled versus three sheets processed simultaneously each one of 2 mm thickness with an inner radius of 1.5 mm

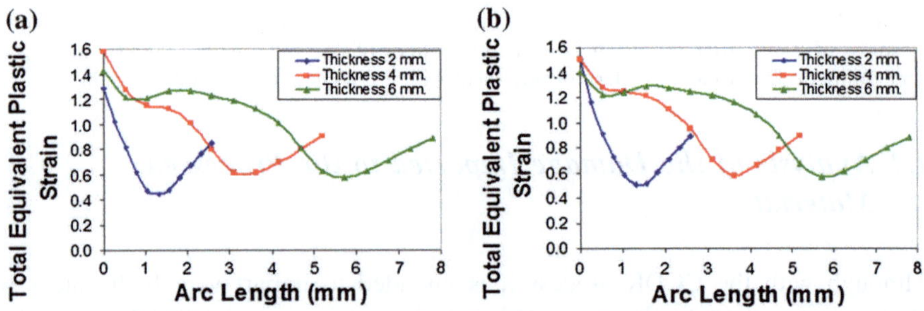

Fig. 14 Plastic deformation in the sections of the sheets with different thickness. **a** Deformation pattern of one sheet of 2 mm thickness after being rolled. **b** Deformation pattern of one sheet of 4 mm thickness after being rolled

been proposed in order to determine the damage mechanics [60–64]. All these formulations have been classified in the three main approaches in [65]: abrupt failure criteria, porous solid plasticity and continuum damage mechanics (CDM).

The first approach predicts the failure when one external variable, that is uncoupled from other internal variables, reaches its critical value. For the second approach, Gurson's type model was derived for porous materials taking into account only the nucleation and growth of the microvoids [60]. This formulation was extended later by other authors [66, 67]. In the last years, as is analysed in [65], a number of finite element unit cell based on micromechanical studies have been performed in order to correlate voids evolution and interaction with the resulting macroscale material yield function. The third approach assumes that the damage is one of the internal constitutive variables that accounts for the effects of the material constitutive response induced by the irreversible processes that occurs in the material microstructure [65]. The initial work in this field was developed by [61]. Subsequently, new models were developed such as those of: Lemaitre [62] and Chaboche [68]. In the las two decades a great number of CMD based formulations have been proposed as can be seen in the revision of Bonora [65]. The limitations of these CMD approaches were overcome with a new damage model formulation proposed in [69].

Although a large number of damage models exist, in the present work, the damage values in the CCDR process have been obtained by using Cockroft–Latham damage models with the Marc Mentat™ FEM software. This model is commonly used in SPD processes [70]. This will allow us to determine the most appropriate die geometry to perform the process. In the present section, the results of the FEM simulations for the damage exerted on the material are shown, as was mentioned, employing a model type Cockroft–Latham [56]. This model can predict a zone of located damage through Eq. (1). Cockroft and Latham damage model is a damage indicator which is often used to predict the initiation of the cracks. It considers the effects of the maximum principal tensile stress. The damage in Cockroft–Latham model is calculated by using Eq. (1).

$$C = \int_0^{\bar{\varepsilon}_f} \frac{\sigma_{\max}}{\bar{\sigma}} d\bar{\varepsilon} \tag{1}$$

where $\bar{\varepsilon}$ represents the equivalent plastic strain, $\bar{\varepsilon}_f$ is the equivalent plastic strain just at the moment of fracture, $\sigma_{\max}$ is the maximum principal tensile stress and C is material constant threshold for damage. As can be observed, no material parameters have to be employed in order to compare how the die geometries influence the damage, when Cockroft–Latham damage model is used. In order to determine how the die geometry influences the damage on the processed material, different FEM models with the above mentioned conditions were simulated in order to calculate the Cockroft–Latham damage parameter. The damage imparted to the material has been evaluated as a consequence of both the rolling process and the angular channel (Fig. 15).

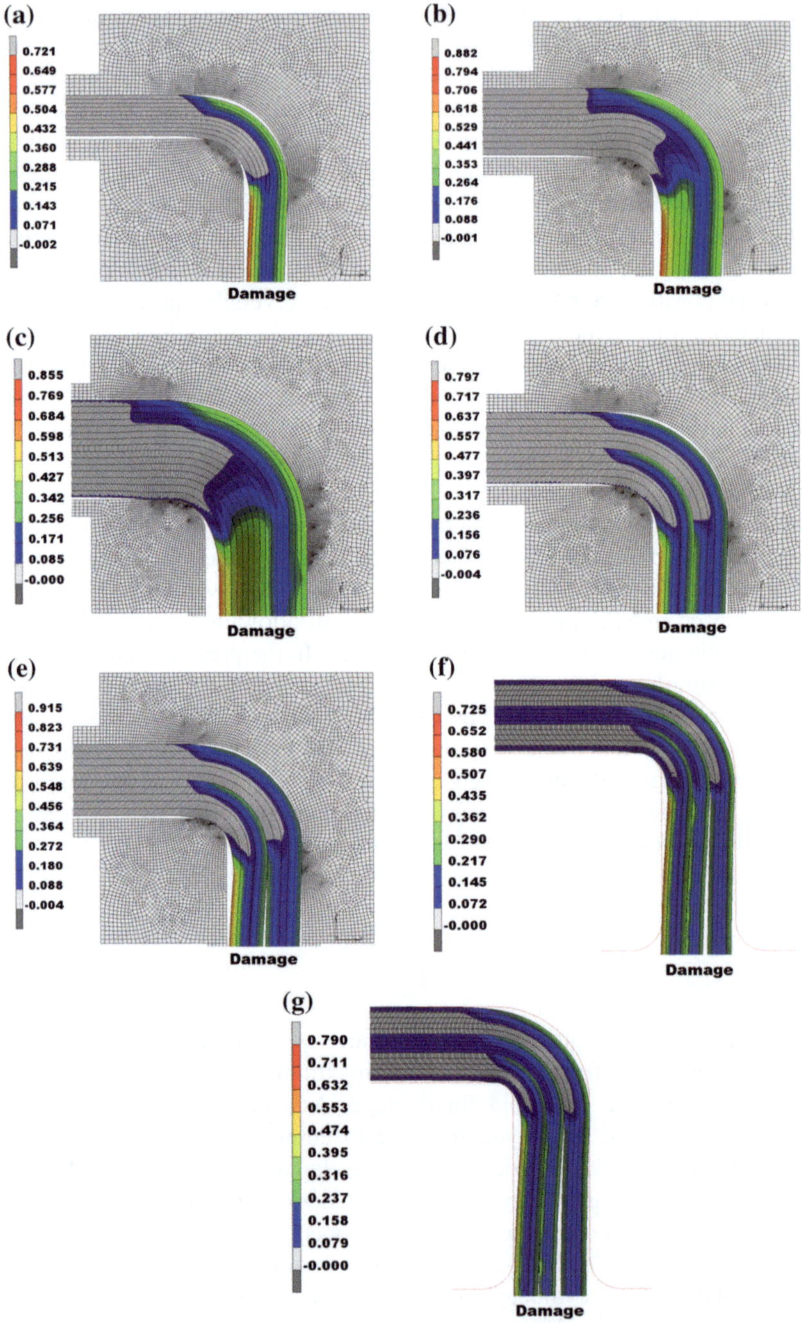
(a)
0.721
0.649
0.577
0.504
0.432
0.360
0.288
0.215
0.143
0.071
-0.002
Damage
(b)
0.882
0.794
0.706
0.618
0.529
0.441
0.353
0.264
0.176
0.088
-0.001
Damage
(c)
0.855
0.769
0.684
0.598
0.513
0.427
0.342
0.256
0.171
0.085
-0.000
Damage
(d)
0.797
0.717
0.637
0.557
0.477
0.397
0.317
0.236
0.156
0.076
-0.004
Damage
(e)
0.915
0.823
0.731
0.639
0.548
0.456
0.364
0.272
0.180
0.088
-0.004
Damage
(f)
0.725
0.652
0.580
0.507
0.435
0.362
0.290
0.217
0.145
0.072
-0.000
Damage
(g)
0.790
0.711
0.632
0.553
0.474
0.395
0.316
0.237
0.158
0.079
-0.000
Damage

◀ **Fig. 15** Damage exerted on the material. **a** CCDR simulation (increment 1,500) in a die with an inner radius of 2 mm and one sheet with a thickness of 2 mm after rolling. **b** CCDR simulation (increment 1,500) in a die with an inner radius of 2 mm and one sheet with a thickness of 4 mm after rolling. **c** CCDR simulation (increment 1,500) in a die with an inner radius of 2 mm and one sheet with a thickness of 6 mm after rolling. **d** CCDR simulation (increment 1,500) in a die with an inner radius of 2 mm and two sheets with a thickness of 2 mm after rolling. **e** CCDR simulation (increment 1,500) in a die with an inner radius of 1.5 mm and two sheets with a thickness of 2 mm after rolling. (f) CCDR simulation (increment 1,650) in a die with an inner radius of 2 mm and three sheets with a thickness of 2 mm after rolling. **g** CCDR simulation (increment 1,650) in a die with an inner radius of 1.5 mm and three sheets with a thickness of 2 mm after rolling

In this section, the effect of the inner radius on the damage imparted to the sheet is shown. The values of the die inner radius (Rint) for the cases studied are 2 and 1.5 mm.

Figure 16a shows the case of a sheet of 2.6 mm thickness rolled up to 2 mm. As can be observed, the model predicts that the damage exerted with a lower inner radius is always higher than that with a higher inner radius. The maximum value is given in the zone closer to the inner radius, followed by a decrease in the central zone and a final increase at the other end of the sheet.

In the zone close to the outer radius, the level of damage is similar in both cases but in the inner part of the sheet, the difference is remarkable, where with a radius of 1.5 mm, the damage is 33 % higher than in the case with an inner radius of 2 mm. Figure 16b shows that the difference between the damage exerted by the different values of radius is lower.

The maximum value is obtained for a radius of 2 mm but the intermediate zone of the sheet is more damaged with a lower value of radius. Figure 16c shows that there is no difference between the damage exerted on the inner zone and on the outer. When the inner radius decreases, the imparted damage is higher, specially for low values of sheet thickness. Moreover, when the sheet thickness is increased, the influence of the inner radius decreases.

In Fig. 17a and b, it can be observed that the damage when a single sheet is processed is higher than that for the two equivalent sheets in section, except in the contact zone of the two sheets, because of the border effect produced when both of them are processed simultaneously. Figure 17c and d show that the simultaneously processed sheets have values of damage lower than those attained in the single processed sheet. In the contact zone between the intermediate sheet and the outer, the level of damage exerted by the friction in these sheets exceeds the sheet with the highest value of thickness.

In the following graphs, Fig. 18, the values of damage for sheets with a starting thickness of 2.6, 5.2 and 7.8 mm, respectively, are shown. One can see that when the thickness is increased, the mean value of damage in the section increases.

The maximum value is achieved for the sheet with the intermediate value of thickness. In this case, it is interesting to employ sheets with a lower value of thickness in order to reduce the level of damage in the central part of the sheet. This result coincides with that obtained in the previous section, where it is shown

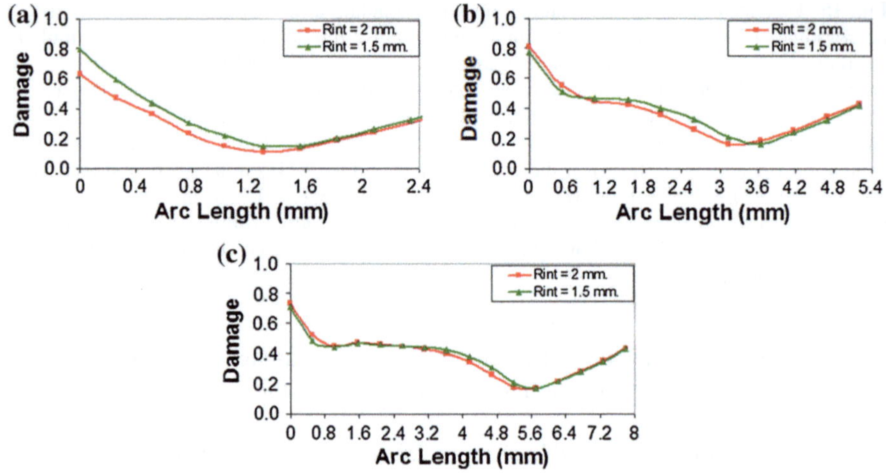

Fig. 16 Values of damage with Cockroft–Latham model in one section of the sheet. **a** Damage pattern for one sheet of 2 mm thickness after being rolled. **b** Damage pattern for one sheet of 4 mm thickness after being rolled. **c** Damage pattern for one sheet of 6 mm thickness after being rolled

that it is preferable to process several sheets of low thickness rather than a single sheet with a higher thickness.

4.3 Equivalent Von Mises Stress Within the Dies

In the present section, the stress distribution inside the CCDR dies is shown when the material is processed with the process conditions which are shown in Table 1. Figure 19 shows the nodes that have been analysed: node 1 corresponds with the inner radius, node 2 corresponds to the contact zone at the upper die before starting the angular channel and node 3 corresponds to the contact zone at the upper die after passing through the angular channel. It has been observed during the execution of the simulations that it is in these points where the sheet makes contact due to a lack of complete filling of the channel in the die.

As can be observed in Fig. 20, the point with the highest level of stress is that at the inner radius of the die, followed by the contact zone after passing through the intersection between both channels. The zone with the lowest value of stress is that at the upper part before the intersection zone between both channels since a big amount of the sheet surface rests over this.

In the image shown in Fig. 21, one can observe the von Mises stress inside the sheets when these are processed by CCDR. The obtained value is that from the material hardening law at the value of plastic deformation which it has accumulated at that moment.

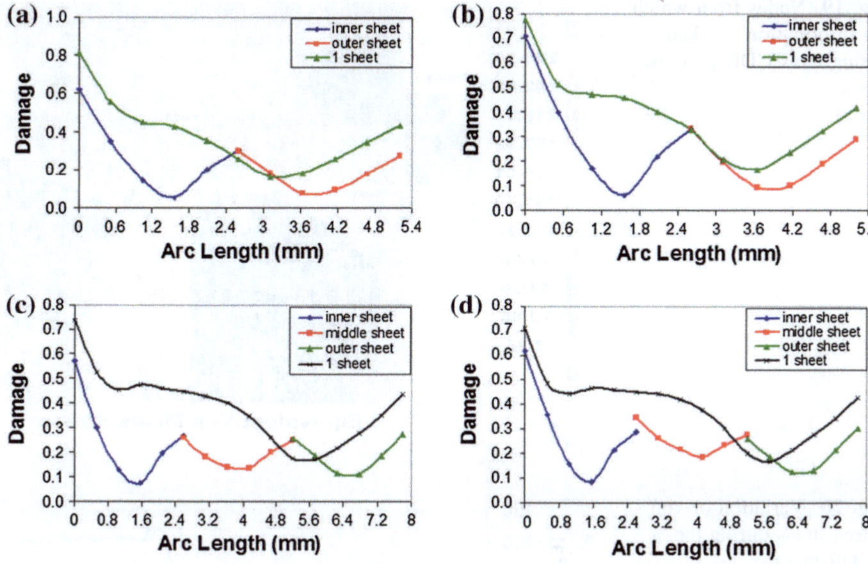

Fig. 17 Values of damage with Cockroft–Latham model in the sections of the sheets. **a** Damage pattern of one sheet of 4 mm thickness after being rolled versus two sheets processed simultaneously each one of 2 mm thickness with an inner radius of 2 mm. **b** Damage pattern of one sheet of 4 mm thickness after being rolled versus two sheets processed simultaneously each one of 2 mm thickness with an inner radius of 1.5 mm. **c** Damage pattern of one sheet of 6 mm thickness after being rolled versus three sheets processed simultaneously each one of 2 mm thickness with an inner radius of 2 mm. **d** Damage pattern of one sheet of 6 mm thickness after being rolled versus three sheets processed simultaneously each one of 2 mm thickness with an inner radius of 1.5 mm

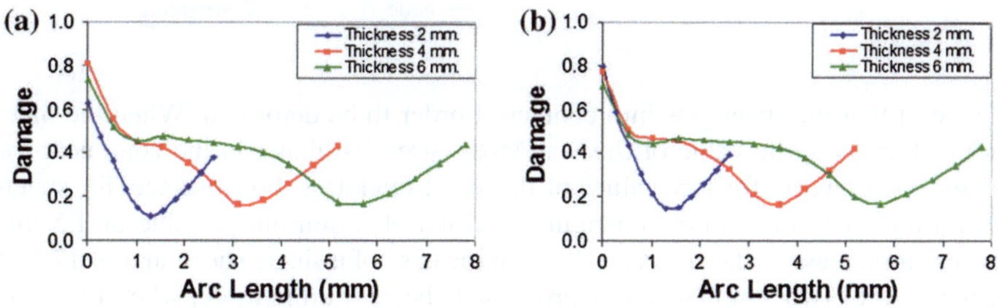

Fig. 18 Values of damage with Cockroft–Latham model in the sections of the sheets with different thickness. **a** Damage pattern for a single sheet with a thickness of 2 mm after being rolled **b** Damage pattern for a single sheet with a thickness of 4 mm after being rolled

Table 2 presents the results obtained from all the simulations which were carried out. It can be concluded that the most affected point when varying the inner radius is node 1, which, in general, presents the highest value of stress since it is at

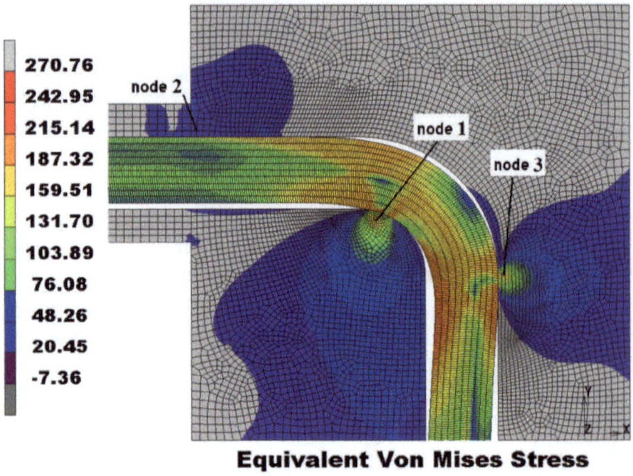

Fig. 19 Nodes from which the information is taken during the CCDR process

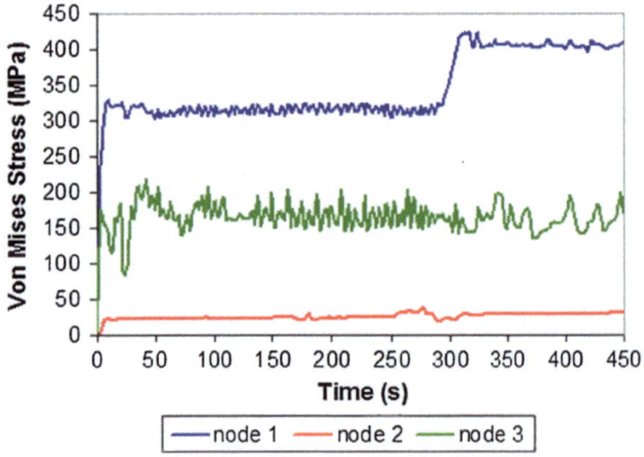

Fig. 20 Variation of von Mises stress during the CCDR process

this point that the sheet gets into contact in order to be deformed. When the inner radius decreases, the value of the von Mises stress is higher in the zone near the node 1, particularly for low values of the sheet thickness. For instance, for sheets with a thickness of 2.6 mm, changing a radius of 2 mm into a value of 1.5 mm means an increase in the stress of 10 %, in the case of a single sheet, and of 17.5 % in the case of two simultaneously processed sheets. Furthermore, when the sheet thickness is increased, the influence of the inner radius diminishes. Figure 22 shows a comparison between the values of stress that appear when a single sheet of 5.2 mm thickness is processed and those attained when two sheets with a thickness of 2.6 mm are processed.

As can be observed in Fig. 22, the stress variation inside the die at nodes 1 and 2 is negligible, whereas in the case of node 3, which corresponds with the outer part of the die, a considerable increase in the stress value is produced. The extreme case is given in the dies with an inner radius of 1.5 mm, where a stress peak

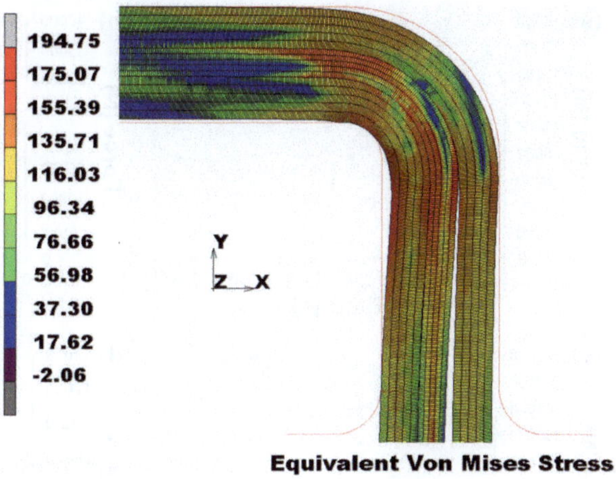

Fig. 21 Von Mises stress inside the sheets during the CCDR process

Table 2 Results of von Mises stress obtained from the different cases considered

Simulations	Node 1	Node 2	Node 3
1 sheet_2.6_2_2.2_90 %_rint2	297.1	30.5	279.4
1 sheet_2.6_2_2.2_90 %_rint1.5	337.8	39.4	254.7
1 sheet_5.2_4_4.2_90 %_rint2	423.1	38.6	216.7
1 sheet_5.2_4_4.2_90 %_rint1.5	427.1	77.0	172.4
1 sheet_7.8_6_6.2_90 %_rint2	410.0	52.2	112.3
1 sheet_7.8_6_6.2_90 %_rint1.5	426.8	56.1	107.3
2 sheets_5.2_4_4.2_90 %_rint2	404.2	24.3	234.5
2 sheets_5.2_4_4.2_90 %_rint1.5	470.1	23.0	392.3

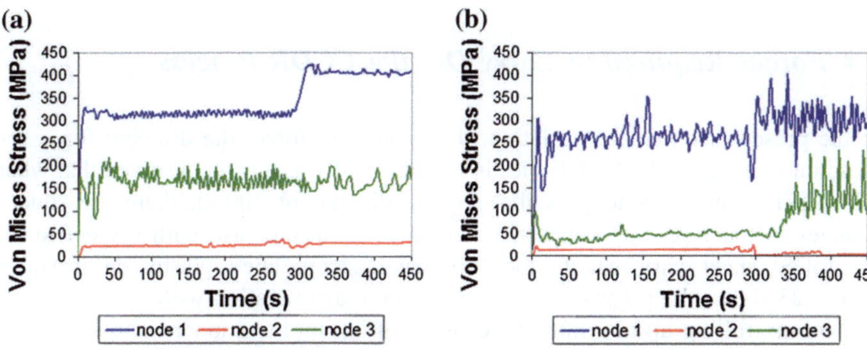

Fig. 22 Von Mises stress inside the sheets during the CCDR process. **a** Single sheet of 2 mm thickness after being rolled with a die of 2 mm inner radius. **b** Two sheets of 4 mm total thickness after being rolled with a die of 1.5 mm inner radius

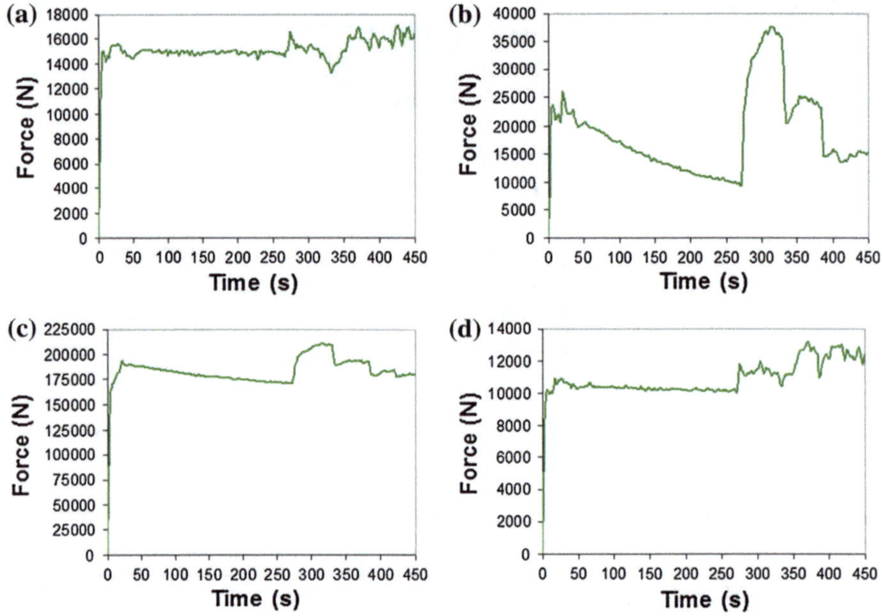

Fig. 23 Diverse forces involved in the CCDR process. **a** Force undergone by the lower die. **b** Force undergone by the upper die. **c** Pushing force carried out by a rolling mill. **d** Drawing force

appears doubling the value for the processing of a single sheet. This result may be due to the fact that with two sheets, a more efficient filling of the channel is produced.

4.4 Forces Required to Carry Out the CCDR Process

In the present section, the variation of the pushing force, the drawing force and those undergone by the CCDR dies is shown. In all cases, it is observed that when the rolled sheet starts to pass through the die (about 300 s), there are sudden changes in the forces and subsequently, there is an increase with respect to the non-rolled sheet part, which does not have accumulated plastic deformation. Figure 23 shows the graphs which correspond to a simulation with an initial sheet of 2.6 mm, rolling up to 2 mm, die channel of 2.2 mm and inner radius of 2 mm.

The force with the highest value involved in the CCDR process is the pushing force, which has to be multiplied by two since it represents the force of only one of the two rolls in the rolling mill.

The drawing force is the lowest of all and its value varies from 5 to 10 % of the pushing force. The values of the force undergone by the dies are counteracted with

Table 3 Forces required to carried out the CCDR process [N]

Simulations	Lower die	Upper die	Pushing force	Drawing force
1 sheet_2.6_2_2.2_90 %_rint2	17,057	37,553	210,285	13,182
1 sheet_2.6_2_2.2_90 %_rint1.5	21,959	41,388	215,981	14,712
1 sheet_5.2_4_4.2_90 %_rint2	51,889	53,141	256,096	31,191
1 sheet_5.2_4_4.2_90 %_rint1.5	50,805	52,668	260,135	31,177
1 sheet_7.8_6_6.2_90 %_rint2	77,131	59,680	292,586	56,998
1 sheet_7.8_6_6.2_90 %_rint1.5	76,402	59,349	293,456	58,112
2 sheets_5.2_4_4.2_90 %_rint2	25,388	17,118	222,796	19,640
2 sheets_5.2_4_4.2_90 %_rint1.5	30,503	22,840	229,451	21,261
3 sheets_7.8_6_6.2_90 %_rint2	43,459	26,788	237,727	37,530
3 sheets_7.8_6_6.2_90 %_rint1.5	43,542	27,250	239,217	39,281

the actual mechanical or hydraulic fasteners which allow us to compensate the force exerted by the material when this is deformed. The obtained results are presented in Table 3.

When the inner radius of the die is varied, no significant changes in the process forces are observed. For instance, when the value of the inner radius is reduced up to 1.5 mm, there is an increase of 15 % in the die forces, in the case of a single sheet with an initial thickness of 2.6 mm and in the case of two sheets with the previously mentioned value of thickness. The values of the force in the lower die decrease significantly when two sheets are processed in relation to those obtained when a single sheet with an equivalent thickness is processed, as is shown in Table 3. The same behaviour can be observed with the drawing force, although the decrease is lower. With respect to the pushing force, it is approximately the same.

From the results of the simulations, it can be stated that the values of the force when three sheets are processed by CCDR are the same as those obtained when a single sheet with a thickness equal to the three of them is processed. It is obvious that in order to reduce the values of the forces, it is better to process the three sheets at the same time. In this case, the clamping force of the dies is reduced by half, whereas the pushing force is reduced an average value of 15 % and the drawing force a 35 %.

The values of the drawing force and of the force undergone by the lower die are doubled when the values for the thickness of the processed sheet are doubled as well, that is to say, they follow approximately a linear law with the thickness. Nevertheless, the pushing force and the force undergone by the upper die increase approximately a 25 % when the sheet thickness is doubled.

5 Conclusions

In the present work, the results obtained from a study about a process denominated as Continuous Combined Drawing and Rolling (CCDR) have been shown. This study has been made by means of the finite element method and employing different process conditions and different types of materials. This process is a new concept of severe plastic deformation processes (SPD), developed by researchers belonging to the Public University of Navarre and based on the patent (ES 2224787) [1]. With the results presented in this work, it is shown that this severe plastic deformation process is industrially viable and it is capable of operating in a continuos form, making it possible to manufacture highly deformed material.

Within the experimental section, it has been seen that the achievement of an optimum value of reduction in the rolling process is a determinant factor for the improvement in the mechanical properties. If the reduction is too high, the filling of the CCDR die is worse and the attained deformation is lower and less homogeneous, which limits the improvement in the mechanical properties of the material.

The decrease in the die inner radius, the processing of a single sheet and the increase in the thickness of the latter increase the deformation introduced in the material. As the improvement in the mechanical properties is linked to the plastic deformation introduced, the adequate selection of the angular channel die geometry will enable a greater improvement in the mechanical properties at each CCDR passage.

When the inner radius is decreased, the imparted damage is higher, especially for low values of sheet thickness. Another alternative in order to achieve a decrease in the level of damage is to process several sheets at the same time, since the mean value of damage in the section increases when the thickness of the processed sheet is increased.

The forces involved in the process increase their values when the inner radius is reduced and the sheet thickness is increased. Nevertheless, as it is the force carried out by the rolls of the rolling mill which is the highest from those involved in the CCDR process, it would be recommendable to increase the sheet thickness because this force increases in a much lower proportion than thickness.

Acknowledgments The authors acknowledge the support given by the Spanish Ministry of Science and Innovation Ministerio de Ciencia e Innovación Project No. MAT 2006-14341-C02-02.

References

1. Luis Pérez, C.J., González. PA., Gil, J., Alkorta, J.: Procesado continuo de materiales metálicos mediante deformación plástica en canal poliangular. Pat. 2224787 (2002)
2. González, P.A., Luis Pérez, C.J., Garcés, Y., Gil Sevillano, J.: ECAE, una tecnología de procesado emergente para producir propiedades relevantes en materiales metálicos. Rev. Metal. **37**, 673–692 (2001)
3. Valiev, R.Z., Islamgaliev, R.K., Alexandrov, I.V.: Bulk nanostructured materials from severe plastic deformation. Prog. Mater. Sci. **45**, 103–189 (2000)
4. Yu, C.Y., Sun, P.L., Kao, P.W., Chang, C.P.: Mechanical properties of submicron-grained aluminum. Scr. Mater. **52**, 359–363 (2005)
5. Roven, H.J., Nesboe, H., Werenskiold, J.C., Seibert, T.: Mechanical properties of aluminium alloys processed by SPD: comparison of different alloy systems and possible product areas. Mater. Sci. Eng. A **410–411**, 426–429 (2005)

6. Xu, C., Furukawa, M., Horita, Z., Langdon, T.G.: Severe plastic deformation as a processing tool for developing superplastic metals. J. Alloy. Compd. **378**(1–2), 27–34 (2004)
7. Chen, H.H., Wang, J.Y., Lee, J., Lee, S.: Superplasticity of AA5083 alloy as processed by equal channel angular extrusión. J. Alloy. Compd. **460**, 305–308 (2008)
8. Segal, V.M.: Equal channel angular extrusion: from macromechanics to structure formation. Mater. Sci. Eng. A **271**, 322–333 (1999)
9. Luis Pérez, C.J.: On the correct selection of the channel die in ECAP processes. Scr. Mater. **50**, 387–393 (2004)
10. Luri, R., Luis, C.J.: Estudio por elementos finitos del proceso de extrusión en canal angular. 1er Congreso Internacional de la Sociedad de Ingeniería de Fabricación (CISIF 05), **1**, 1–8 (2005)
11. Luri, R., León, J., Luis, C.J., Puertas, I.: Mechanical behaviour of an Al–Mg alloy processed by ECAE. In: 21st International Manufacturing Conference (IMC 21), vol. 1, pp. 167–174 (2004)
12. Semiatin, S.L., Delo, D.P.: Equal channel angular extrusion of difficult-to work alloys. Mater. Des. **21**, 311–322 (2000)
13. Kim, H.S., Seo, M.H., Hong, S.I.: On the die corner gap formation in equal channel angular pressing. Mater. Sci. Eng. A **291**, 86–90 (2000)
14. Bowen, J.R., Gholinia, A., Roberts, S.M.: Analysis of the billet deformation behaviour in equal channel angular extrusión. Mater. Sci. Eng. A **287**, 87–99 (2000)
15. Tsai, T.L., Sun, P.L., Kao, P.W., Chang, C.P.: Microstructure and tensile properties of a commercial 5052 aluminum alloy processed by equal channel angular extrusión. Mater. Sci. Eng. A **342**, 144–151 (2003)
16. Sun, P.L., Cerreta, E.K., Gray, G.T., Rae, P.: The influence of boundary structure on the mechanical properties of ultrafine grained AA1050. Mater. Sci. Eng. A **410–411**, 265–268 (2005)
17. Wang, Z.C., Prangnell, P.B.: Microstructure refinement and mechanical properties of severely deformed Al–Mg–Li alloys. Mater. Sci. Eng. A **328**, 87–97 (2002)
18. Alkorta, J., Luis Pérez, C.J., Popova, E.N., Hafok, M., Pippan, R., Gil Sevillano, J.: Microstructure and indentation size-effect in pure niobium subjected to SPD via ECAP and HPT. Mater. Sci. Forum. **584–586**, 215–220 (2008)
19. Ferrasse, S., Segal, V.M., Alford, F., Kardokus, J., Strothers, S.: Scale up and Aplication of Equal Channel Angular Extrusion (ECAE) for the Electronics and Aerospace Industries. Mater. Sci. Eng. A **493**, 130–140 (2008)
20. Garcés, Y., Luis, C.J., Berlanga, C., González, P.: Equal channel angular extrusion in a commercial Al–Mn alloy. In: International Conference in Advances in Materials and Processing Technologies (AMPT'01), vol. 1, pp. 267–274 (2001)
21. González, P.A., Luis, C.J.: ECAE analysis of a 5083 Aluminiun–Magnesium alloy. Materials Congress (2002)
22. Lugo, N., Cabrera, J.M., Llorca, N., Luis, C.J., Luri, R., León, J., Puertas, I.: Grain refinement of pure copper by ECAP. Mater. Sci. Forum. **584–586**, 393–398 (2008)
23. Kim, K.J., Yang, D.Y., Yoon, J.W.: Investigation of microstructure characteristics of commercially pure aluminum during equal channel angular extrusión. Mater. Sci. Eng. A doi:10.1016/j.msea.2007.08.038 (2007)
24. Khan, Z.A., Chakkingal, U., Venugopal, P.: Analysis of forming loads, microstructure development and mechanical property evolution during equal channel angular extrusion of a commercial grade aluminum alloy. J. Mater. Process. Technol. **135**, 59–67 (2003)
25. Llorca Isern, N., Gonzalez, P.A., Luis, C.J., Laborde, I.: Severe plastic deformation of a commercial aluminium-lithium alloy (AA8090) by Equal Channel Angular Pressing. Mater. Sci. Forum. **503–504**, 871–876 (2006)
26. Segal, V.M.: Materials processing by simple shear. Mater Sci Eng. **197**, 157–164 (1995)
27. Lapovok, R., Loader, C., Dalla Torre, F.H., Semiatin, S.L.: Microstructure evolution and fatigue behavior of 2124 aluminum processed by ECAE with back pressure. Mater. Sci. Eng. A **425**, 36–46 (2006)

28. González PA., Luis CJ (2002) Severe Plastic Deformation by ECAP in an Al-Mg-Mn Commercial Alloy. 2nd Conference on SPD, NanoSPD2 1:251-256
29. Humphreys, F.J., Prangnell, P.B., Priestner, R.: Fine-grained alloys by thermomechanical processing. Solid State Mater. Sci. **5**, 15–21 (2001)
30. Huarte, B., Luis, C.J., Puertas, I., León, J., Luri, R.: Optical and mechanical properties of an Al–Mg alloy processed by ECAE. J. Mater. Process. Technol. **162–163**, 317–326 (2005)
31. Nakashima, K., Horita, Z., Nemoto, M., Langdon, T.G.: Development of multi-pass for equal-channel angular pressing to high total strains. Mater. Sci. Eng. A **281**, 82–87 (2000)
32. Neishi, K., Horita, Z., Langdon, T.G.: Grain refinement of pure nickel using equal-channel angular pressing. Mater. Sci. Eng. A **325**, 54–58 (2002)
33. Janecek, M., Popov, M., Krieger, M.G., Hellming, R.J., Estrin, Y.: Mechanical properties and microstructure of a Mg alloy AZ31 prepared by equal-channel angular pressing. Mater. Sci. Eng. A **462**, 116–120 (2007)
34. Shin, D.H., Seo, C.W., Kim, J., Park, K., Choo, W.Y.: Microstructures and mechanical properties of equal-channel angular pressed low carbon steel. Scr. Mater. **42**, 695–699 (2000)
35. Kaibyshev, R., Shipilova, K., Musin, F., Motohashi, Y.: Continuous dynamic recrystallization in an Al–Li–Mg–Sc alloy during equal-channel angular extrusión. Mater. Sci. Eng. A **396**, 341–351 (2005)
36. Nieh, T.G., Wadsworth, J., Sherby, O.D.: Superplasticity in Metals and Ceramics. Cambridge University Press, London (1997)
37. Li, C., Xia, Z., Sue, H.: Simple shear plastic deformation of polycarbonate plate II. Mechanical property characterization. Polymer **41**, 6285–6293 (2000)
38. Korbel, A., Richert, M.: Formation of shear bands during cyclic deformation of aluminium. Acta Metall. **33**, 1971–1978 (1985)
39. Richert, M., Stüwe, H.P., Zehetbauer, M.J., Richert, J., Pippan, R., Motz, Ch., Schafler, E.: Work hardening and microstructure of AlMg5 after severe plastic deformation by cyclic extrusion and compression. Mater. Sci. Eng. A **355**, 180–185 (2003)
40. Wang, Q.D., Chen, Y.J., Lin, J.B., Zhang, L.J., Zhai, C.Q.: Microstructure and properties of magnesium alloy processed by a new severe plastic deformation. Mater. Lett. **61**, 4599–4602 (2007)
41. Akbari Mousavi, S.A., Shahab, A.R., Mastoori, M.: Computational study of Ti-6Al-4 V flow behaviors during the twist extrusion process. Mater. Des. **29**, 1316–1329 (2008)
42. Vorhauer, A., Pippan, R.: On the homogeneity of deformation by high pressure torsion. Scr. Mater. **51**, 921–925 (2004)
43. Jiang, H., Zhu, Y.T., Butt, D.P., Alexandrov, I.V., Lowe, T.C.: Microstructural evolution, microhardeness and thermal stability of HPT-processed Cu. Mater. Sci. Eng. A **290**, 128–138 (2000)
44. Sakai, G., Horita, Z., Langdon, T.G.: Grain refinement and superplasticity in an aluminum alloy processed by high-pressure torsión. Mater. Sci. Eng. **393**, 344–351 (2005)
45. Stolyarov, V.V., Zhu, Y.T., Lowe, T.C., Islamgaliev, R.K., Valiev, R.Z.: A two step SPD processing of ultrafine-grained titanium. Nanostruct. Mater. **11**, 947–954 (1999)
46. Sergueeva, A.V., Stolyarov, V.V., Valiev, R.Z., Mukherjee, A.K.: Advanced mechanical properties of pure titanium with ultrafine grained structure. Scr. Mater. **45**, 747–752 (2001)
47. Rajinikanth, V., Arora, G., Narasaiah, N., Venkateswarlu, K.: Effect of repetitive corrugation straightening on Al and Al-0.25 Sc alloy. Mater. Lett. **62**, 301–304 (2008)
48. Huang, J., Zhu, Y.T., Alexander, D.J., Liao, X., Lowe, T.C., Asaro, R.J.: Development of repetitive corrugation and straightening. Mater. Sci. Eng. A **371**, 35–39 (2004)
49. Ghosh, A.K., Huang, W.: Severe deformation based process for grain subdivision and resulting microstructures. Invest. appl. Severe Plast. Deformation **1**, 29–36 (2000)
50. Valiev, R.Z., Langdon, T.G.: Principles of equal-channel angular pressing as a processing tool for grain refinement. Prog. Mater. Sci. **51**, 881–981 (2006)
51. Zhu, Y.T., Lowe, T.C., Langdon, T.G.: Performance and applications of nanostructured materials produced by severe plastic deformation. Scr. Mater. **51**, 825–830 (2004)

52. Latysh, V., Krallics, G.Y., Alexandrov, I., Fodor, A.: Application of bulk nanostructured materials in medicine. Curr. Appl. Phys. **6**, 262–266 (2006)
53. Tanaka, T., Makii, K., Ueda, H., Kushibe, A., Kohzu, M., Higashi, K.: Study on practical application of a new seismic damper using a Zn–Al alloy with a nanocrystalline microstructure. Int. J. Mech. Sci. **45**, 1599–1612 (2003)
54. Luis Pérez, C.J., León Iriarte, J.: Fem simulation of the continuous combined drawing and rolling pressing in equal channel angular (CCDR-ECAP). UFG3 **1**, 205–210 (2004)
55. Luis Pérez, C.J., León Iriarte, J., Díaz de Rada, C.: A method for continuous processing of materials by equal channel angular extrusion processes. In: International Conference on Advanced Materials and Processing Technologies, AMPT'03, vol. 1, pp. 1162–1165 (2003)
56. Cockroft, M.G., Latham, D.J.: Ductility and the workability of metals. Inst. Met. **96**, 33–39 (1968)
57. Marc.: R1, Theory and User information A (2008)
58. Lemaitre, J.: A Course on Damage Mechanics. Springer, New York (1996)
59. Bonora, N., Ruggiero, A., et al.: Practical applicability and limitations of the elastic modulus degradation technique for damage measurements in ductile metals. Strain (2010). doi:10.1111/ j.1475-1305.2009.00678.x
60. Gurson, A.L.: Continuum theory of ductile rupture by void mucleation and growth: part I—Yield criteria and flow rules for porous ductile media. J. Eng. Mat. Tech. **99**, 2–15 (1977)
61. Kachanov, L.M.: On creep rupture time. Izv. Acad. Nauk SSSR, Otd. Techn. Nauk **88**, 26–31 (1958)
62. Lemaitre, J.: A continuous damage mechanics model for ductile fracture. J. Eng. Mat. Tech. **107**, 83–89 (1985)
63. Murakami, S.: Damage mechanics approach to damage and fracture of materials. Rairo **3**, 1–13 (1982)
64. Krajcinovic, D.: Damage mechanics. Mech. Mat. **8**, 117–197 (1989)
65. Bonora, N., Gentile, D., et al.: Ductile damage evolution under triaxial state of stress: theory and experiments. Int. J. Plast. **21**, 981–1007 (2005)
66. Needleman, A., Tvergaard, V.: An analysis of ductile ruptura in biaxially stretched sheets. J. Eng. Mat. Tech. **102**, 249–256 (1984)
67. Gao, X., Faleskog, J., Shih, C.F., Dodds, R.H.J.: Ductile tearing in part-through cracks: experiments and cell-model predictions. Eng. Fract. Mech. **59**, 761–777 (1998)
68. Chaboche, J.L.: Anisotropic creep damage in the framework of the continuum damage mechanics. Nucl. Eng. Des. **79**, 309–319 (1984)
69. Bonora, N.: A non-linear CMD damage model for ductile failure. Eng. Fract. Mech. **58**, 11–28 (1997)
70. Luri, R., Luis Pérez, C.J., et al.: Evolution of damage in AA-5083 processed by equal cannel angular extrusión using different die geometries. J. Mater. Process. Technol. **211**, 48–56 (2011)

Simulation of Thermal and Electrical Transport in Nanotube and Nanowire Composites

Satish Kumar, Muhammad A. Alam and Jayathi Y. Murthy

Abstract Nanotube-based thin-film composites promise significant improvement over existing technologies in the performance of large-area macroelectronics, flexible electronics, energy harvesting and storage, and in bio-chemical sensing applications. We present an overview of recent research on the electrical and thermal performance of thin-film composites composed of random 2D dispersions of nanotubes in a host matrix. Results from direct simulations of electrical and thermal transport in these composites using a finite volume method are compared to those using an effective medium approximation. The role of contact physics and percolation in influencing electrical and thermal behavior are explored. The effect of heterogeneous networks of semiconducting and metallic tubes on the transport properties of the thin film composites is investigated. Transport through a network of nanotubes is dominated by the interfacial resistance at the contact of two tubes. We explore the interfacial thermal interaction between two carbon nanotubes in a crossed configuration using molecular dynamics simulation and wavelet methods. We pass a high temperature pulse along one of the nanotubes and investigate the energy transfer to the other tube. Wavelet transformations of heat pulses show that how different phonon modes are excited and how they evolve and propagate along the tube axis depending on its chirality.

S. Kumar (✉)
G. W. Woodruff School of Mechanical Engineering, Georgia Institute of Technology, Atlanta, GA 30332, USA
e-mail: satish.kumar@me.gatech.edu

M. A. Alam
School of Electrical and Computer Engineering, Purdue University, West Lafayette, IN 47097, USA
e-mail: alam@ecn.purdue.edu

J. Y. Murthy
School of Mechanical Engineering, Purdue University, West Lafayette, IN 47097, USA
e-mail: jmurthy@ecn.purdue.edu

N. Bonora and E. Brown (eds.), *Numerical Modeling of Materials Under Extreme Conditions*, Advanced Structured Materials 35, DOI: 10.1007/978-3-642-54258-9_3,

Keywords Nanotube · Thin film transistor · Nanocomposite · Percolation · Effective medium approximation · Molecular dynamics · Wavelet

1 Introduction

In recent years, there has been enormous interest in fabricating thin-film transistors (TFTs) on flexible substrates in the rapidly growing field of large-area macro-electronics [1, 2]. Applications include displays [1], e-paper, e-clothing, pressure-sensitive skin [3, 4], large-area chemical and biological sensors [5, 6], flexible and shape-conformable antennae and radar, as well as intelligent and responsive surfaces with large-area control of temperature, drag and other properties [2]. Flexible substrates such as plastic require low temperature processing, typically below 200 °C. Prevailing technologies such as amorphous silicon (a-Si) and organic TFTs can be processed at low temperature and are sufficient for low-performance applications such as displays, where their low carrier mobility ($\sim$1–10 cm^2/Vs) [4, 7] is not a limitation. For high-performance applications, however, the choices are limited. Single crystal silicon CMOS and polycrystalline silicon (poly-Si) technologies can yield higher performance, but are expensive and cannot be fabricated below 250 °C. Nanotube-bundle (NTB) based TFTs, consisting of carbon nanotubes (CNTs) dispersed in substrates such as polymer and glass, are being explored to substantially increase the performance of flexible electronics to address *medium-to-high performance* applications in the 10–100 MHz range [2]. High mobility, substrate-neutrality and low-cost processing make NTB-TFTs very promising for these flexible-electronics applications.

Two distinct classes of materials are being pursued by researchers [8–11]. On the one hand, *randomly-oriented* nanotubes embedded in polymer have been used to fabricate nanotube bundle thin-film transistors (NTB-TFTs) which promise relatively high carrier mobility ($\sim$100 cm^2/Vs). Here, solution-processing is used to disperse a random network of CNTs in a plastic substrate, as shown in Fig. 1, to form a thin film. The mat of CNTs forms the channel region of the transistor. Because NTB-TFTs do not require precise alignment of CNTs, they are amenable to mass manufacture, and are relatively inexpensive. Recently, a few groups have begun to fabricate these TFTs for macro-electronics and chemical sensing applications [7, 8] and have begun to explore their performance. Snow et al. reported the mobility and conductance properties of carbon nanotube (CNT) networks and also explored the interfacial properties of CNTs in chemical sensing transistors [9–11]. Menard et al. fabricated thin film transistors on plastic substrates using nano-scale objects (microstrips, platelets, disks, etc.) of single crystal silicon. Zhou et al. demonstrated fabrication of p-type and n-type transistors [8], which could be used as building blocks for complex complementary circuits. Fabrication of an integrated digital circuit composed of up to nearly 100 transistors on plastic substrates using random network of CNTs has been reported by Cao and Rogers [12]. Other

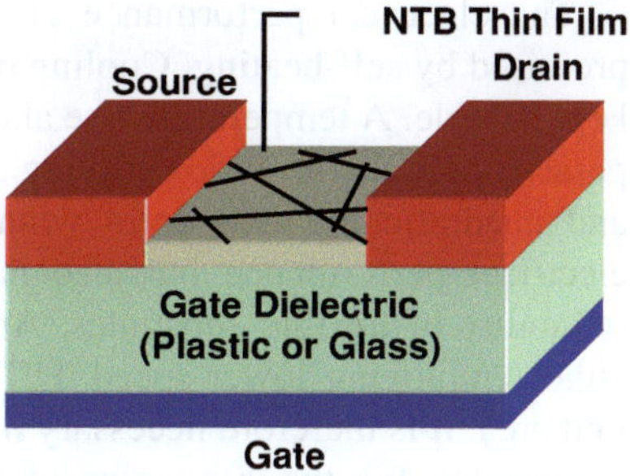

Fig. 1 Schematic of nanotube bundle thin-film transistor showing source, drain, gate and channel region composed of nanotube composite

experimental reports on CNT TFT fabrication can be found in [2, 12, 13]. A number of groups have focused on developing TFTs with *well-aligned* and *or partially-aligned* nanotubes for very high performance applications [14–16] using transfer printing; mobilities of 1,000 cm^2/Vs, comparable to single-crystal silicon are achievable using this technology.

Though there has been a great deal of research on composites, [17, 18] nano-composites for use in macrolectronics pose very specific problems. First, unlike most published research on 3D transport in composites, our interest is in 2D thin-film composites in which in-plane electrical transport dominates, and in which in-plane thermal spreading plays a central role in determining device temperature. Furthermore, macroelectronic devices are typically of the 1–50 micron scale. At these scales, the nanotube length may compete with the finite size of the device, and unlike in most published research, bulk composite behavior does not obtain. Furthermore, there remain a large number of unknowns regarding the ultimate performance limits of NTB-TFTs. For example, nearly all reported work has concentrated on device fabrication and processing, but little is understood about the fundamental physics that govern device operation and scaling as a function of tube orientation, tube density, ratio of metallic to semiconducting tubes, and tube-substrate interaction [19].

Strong electrical, thermal and optical interactions between the tubes and between the tubes and the substrates affect device performance, but there has been little fundamental work to explore these interactions quantitatively. Furthermore, metallic CNTs form 30 % of typical NTBs which are problematic because they can short source and drain and limit on-off ratios [20]. Recently, a number of techniques for removing them have been reported. A gas-phase plasma hydro-carbonation reaction technique has been reported to selectively etch and gasify metallic nanotubes and obtain pure semiconducting nanotubes [21]. Another process that separates single-walled carbon nanotubes (SWNTs) by diameter, band gap and electronic type using centrifugation of compositions has been reported by Ref. [22]. The degree to which metallic tubes can influence on/off ratios must be understood for controlled and optimal design. Last but not least, the supply voltage used thus far in driving these devices has been untenably high, leading to unacceptable power dissipation and hysteresis due to charge injection. Processing conditions must be optimized to reduce the supply voltage to acceptable values.

The electrical performance of NTB-TFT macroelectronics is severely compromised by self-heating. Cooling options are limited if macroelectronics are to be kept flexible. A temperature rise above ambient in the 100 °C range is expected for passive natural convection cooling and is expected to scale linearly with frequency and quadratically with drain voltage. High temperatures not only compromise electrical performance but also have consequences for the thermo-mechanical reliability of flexible substrates. An inability to control self-heating would mean either employing lower-speed TFTs or decreasing the number of transistors per unit area. It is therefore necessary not only to understand thermal transport in these composites, but the interaction of electrical and thermal transport in determining device performance and reliability.

Low density CNT composites have been extensively explored for applications in thermal management [23–25], and high strength materials [26, 27]. In these applications, CNTs are embedded in host substrate as a random matrix. A percolating network of CNTs is found to be formed even at low volume fractions ($\sim$0.2 %) due to their high aspect ratio [25, 28]. Theoretical and numerical studies based on the effective medium approximation (EMA) [18, 24], Monte-Carlo simulations [29] or on scaling analysis [30] have been reported on percolating nanotube networks or their composites to predict their effective electrical or thermal transport properties. However, many of these studies significantly limit the thermal conductivity ratios addressed and do not address finite-sized 2D composites. A number of experimental measurements of effective thermal conductivity (k_{eff}) of nanotube suspensions in either substrates or fluids have been reported recently, and are summarized in Table 1. There are large disparities in the reported enhancement of k_{eff} over that of the substrate. However, all experiments show that the maximum achievable conductivity is less than three times that of the underlying matrix/fluid, a harbinger of thermal problems in NTB-TFTs. It is necessary to understand and control the physics underlying these performance limits, particularly the influence of tube-tube and tube-substrate contact parameters on k_{eff}.

A firm understanding of tube-tube and tube-substrate interfacial transport may provide guidelines for improving the efficiency and reliability of CNT based devices. Various experimental and numerical studies have been performed to estimate the thermal conductivity of CNTs and also to measure the thermal resistance between the CNT and the substrate. Most numerical studies are based on the molecular dynamics (MD) method [31, 32]. A list of these studies may be found in Ref. [31]. Small et al. measured the tube-to-substrate resistance (on a per-length basis) of 12 Km/W for a MWNT sitting on a substrate [33] and Maune et al. determined the thermal resistance between a SWCNT and a solid sapphire substrate as 3 Km/W [34]. Recently, Carlborg et al. studied the thermal boundary resistance and the heat transfer mechanism between CNTs and an argon matrix using MD [35].

Recent molecular dynamics (MD) computations [32, 36] have found high values for tube-tube contact resistance. Maruyama et al. used MD simulations to compute the thermal boundary resistance between a CNT surrounded by six other CNTs using the lumped capacitance method [32]. By measuring the transient

Table 1 Experimental results for enhancement in effective thermal conductivity of CNT composites

Composite type	Diameter (d, nm)	Length (L_t,)	Concentration (%)	Increase in k_{eff} (%) over substrate	Network (type)
SWNT/silicon elastomer [65]	–	–	3.8 wt.%	65	Random
SWNT/epoxy [25]	1.1	–	1 wt.%	125	Random
SWNT/epoxy [56]	1.1	170 nm	1 wt.%	80	Random
MWNT/carbon [66]	20–40	10–100 μm	–	12.31	Aligned
MWNT/silicon elastomer [23]	12	0.3 mm	0.4 vol.%	280	Aligned
MWNT/oil [67]	25	50 μm	1 vol.%	160	Random
MWNT/water [68]	20–60	>10 μm	0.84 vo.1%	27.3	Random

temperature change of CNTs they computed the CNT-CNT thermal resistance, and found it to be of the order of 1.0×10^{-7} m^2-K/W [32]. Zhong and Lukes considered heat transfer between CNTs using classical MD simulations and estimated the interfacial thermal transport between offset parallel single-wall carbon CNTs as a function of CNT spacing, overlap, and length [36]. Greaney and Grossman used MD techniques to understand the effect of resonance on the mechanical energy transfer between CNTs [37]. It has been shown recently that the thermal resistance at a CNT-CNT contact should be of the order of 0.3×10^{-12} K/W to match the very low conductivity measured for CNT beds. This has also been verified by atomistic Green's function (AGF) simulations [38]. Nevertheless, the mechanism of energy transport at the interface of two CNTs is still not well understood and needs a detailed exploration. Experimental techniques for direct measurement of CNT-CNT resistance have not yet been reported; thus atomistic-level simulations are a vital tool with which to analyze the interfacial transport mechanism.

In this article, we present an overview of our recent research on the simulation of the performance of CNT nanocomposites for the macro-electronic applications and thermal interaction between two contacting CNTs. A generalized finite volume approach is presented for evaluating the electrical and thermal conductivity and device performance of nanotube bundle TFTs composed of finite two-dimensional nanocomposites. We first apply the approach to the prediction of the electrical and thermal conductance of pure percolating networks of CNTs in the absence of a substrate. Predictions of the electrical characteristics of pure-network TFTs in the linear regime are then presented and their behavior is explained by invoking the physics of heterogeneous finite-sized networks of metallic and semiconducting tubes. The numerical results for the estimation of the effective conductance properties of composites are presented next which explore the effect of tube-to-tube conductance, tube-to-substrate conductance and network density on both electrical and thermal transport. A two-dimensional effective medium approximation is derived for thin-film composites and compared to our numerical

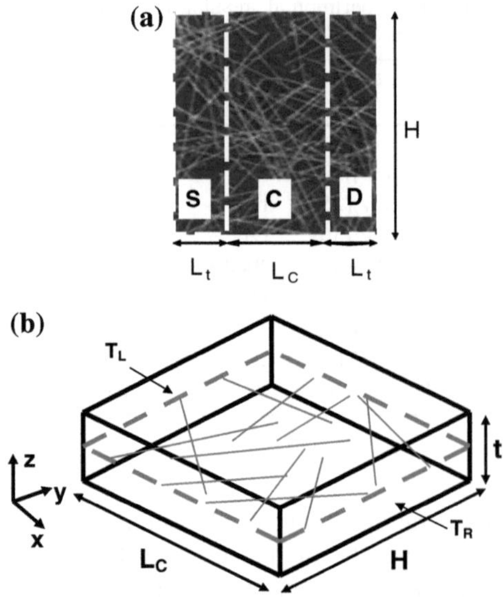

Fig. 2 **a** Schematic of thin-film transistor showing source (*S*), drain (*D*) and channel (*C*). The channel region is composed of a network of CNTs, **b** geometric parameters

simulations, and deficiencies in the EMA model for CNT composites are identified. Attention is turned next to the analysis of the thermal transport physics between two CNTs using molecular dynamics simulations and wavelet methods. We investigate the thermal interaction between two CNTs in a crossed configuration when a high temperature pulse is passed along one of the CNTs. Wavelet analysis decomposes the time series of the heat pulse in the time-frequency space and helps in determining the evolution and propagation of dominant modes.

2 Numerical Formulation

A schematic of a nanotube bundle transistor is shown in Fig. 1. A typical transistor is a four terminal device in which the four terminals are the source (S), drain (D), gate (G) and substrate, Fig 2. TFTs are a special class of transistor in which a thin film of semi-conducting material is used as the channel (C) region between the source and drain. In nanotube bundle TFTs, a thin CNT composite film acts as the channel region. Typical dimensions are indicated in Fig. 2b, where L_C is the length of the channel, L_t is the average length of the nanotubes, d is the diameter of nanotube, H is the width of the transistor and t is the thickness of the nanocomposite. A fixed voltage bias, V_{DS}, is applied across the channel from drain to source to drive the mobile charges in the channel region, while the transistor is turned on and off by changing the gate voltage, V_{GS}. The corresponding current is denoted by I_{DS}. An important parameter to assess device performance is the on-off ratio

(R) which is the ratio of the current flowing in the device in the on-state, I_{ON}, to the current in device in the off-state, I_{OFF} .

In the present article, our analysis of the electrical performance of NTB TFTs is limited to the linear regime, a regime where current (I_{DS}) through the device is linearly proportional to V_{DS}. This is only true at low V_{DS}. A extension of this problem has been reported by Ref. [39] that generalizes this problem to high-bias regime (high V_{DS}) and provides proper scaling laws to predict the performance of transistors with arbitrary geometrical parameters and biasing conditions. For an insulating substrate (either for electrical or thermal transport), only transport in the percolating network of tubes is considered and the effective conductivity/conductance of the pure network is computed (see Sect. 3), Fig. 2a. If the substrate is sufficiently conducting, transport in both substrate and tube network are considered for computing effective conductive properties (see Sect. 4), incorporating the effect of tube-to-substrate interaction.

2.1 Thermal Transport

The computational domain for computing effective thermal properties of the nanotube composite is a three-dimensional box of size L_C x H x t (see Fig. 2b), which is composed of a 2D random network of nanotubes embedded in the mid-plane of the substrate. Diffusive transport in the tube obtains when there are a sufficient number of scattering events during the residence time of the phonon in the tube. This condition prevails here because of the dominance of interface scattering at the tube-substrate boundary. Thus, Fourier conduction in the nanotubes may be assumed, albeit with a thermal conductivity that may differ significantly from bulk or freestanding values. Assuming one-dimensional diffusive transport along the length s of the tube and three-dimensional conduction in the substrate, the governing energy equations [19] in the tube and substrate may be written in non-dimensional form as:

Tube:

$$\frac{d^2\theta_i}{ds^{*2}} + \sum_{\text{intersecting tubes}j} Bi_c(\theta_j - \theta_i) + Bi_s(\theta_s - \theta_i) = 0, \tag{1a}$$

Substrate:

$$\nabla^{*2}\theta_s + \sum_{i=1}^{N_{tubes}} Bi_s\beta_v \frac{k_t}{k_s}(\theta_i - \theta_s) = 0. \tag{1b}$$

Here, dimensionless temperature variable is $\theta = (T-T_L)/(T_R-T_L)$; T_R and T_L are the face temperatures of the right and left boundary faces of the composite (see Fig. 2b). These are the faces which contact source and drain when the thin film composite is used as the channel in the transistor. All lengths are non-dimensionalized by the tube diameter d. $\theta_i(s^*)$ is the non-dimensional temperature of the ith tube at a location s^* along its length and θ_s is the substrate temperature. The other dimensionless parameters are defined as:

$$Bi_c = \frac{h_c P_c d^2}{k_t A}; \quad Bi_s = \frac{h_s P_s d^2}{k_t A}; \quad \frac{k_s}{k_t}; \quad \beta_v = \alpha_v \left(\frac{A}{P_s}\right); \quad \frac{\mathrm{L_C}}{\mathrm{L_t}}; \quad \frac{H}{L_t}; \quad \frac{\mathrm{L_t}}{\mathrm{d}}; \quad \frac{t}{d}$$

Here, Bi_c represents the dimensionless contact conductance for tube-to-tube contact; Bi_s represents the dimensionless interfacial resistance between the tube and substrate, both due to Kapitza resistance and isotherm distortion near the tube. A is the effective cross-section of the tube, and k_t is the corresponding thermal conductivity. The term h_c is the heat transfer coefficient governing the transfer of heat to other tubes j making contact with tube i through a contact perimeter P_c, and the heat transfer coefficient h_s governs the transfer of heat between the tube and the substrate through a contact perimeter P_s. k_s is the substrate thermal conductivity. The second term in Eq. (1b) contains the heat exchange with tubes traversing the substrate, which are $\mathrm{N_{tubes}}$ in number, through a contact area per unit volume, α_v. The geometric parameter β_v may be determined from the tube density per unit area ρ and the corresponding dimensionless parameter is $\rho^*(\rho/\rho_{th})$.The percolation threshold (ρth) for the network is estimated as the density at which the average distance between the nanotubes equals the average length of the tubes, so that $\rho_{th} = 1 \big/ \langle L_t \rangle^2$.

For thermal conductivity calculations, the thermal boundary conditions for all tubes originating at the source and terminating in the drain are given by:

$$\theta_i = 1 \text{ at } \mathrm{s}^* = 0; \quad \theta_i = 0 \text{ at } \mathrm{s}^* = \frac{\mathrm{L_t}}{\mathrm{d}}, \tag{2}$$

and the boundary conditions for the substrate are given by:

$$\theta_s = 1 \text{ at } \mathrm{x}^* = 0; \quad \theta_s = 0 \text{ at } \mathrm{x}^* = \frac{\mathrm{L_C}}{\mathrm{d}}; \quad \frac{\partial \theta_s}{\partial z^*} = 0 \text{ at } \mathrm{z}^* = 0 \quad \text{and} \quad \text{at } \mathrm{z}^* = \frac{\mathrm{t}}{\mathrm{d}}. \tag{3}$$

All the tube tips terminating inside the substrate are assumed adiabatic. The boundaries $\mathrm{y}^* = 0$ and $\mathrm{y}^* = \mathrm{H/d}$ are assumed as periodic boundaries for both substrate and tubes.

2.2 Electrical Transport

The dimensionless potential equation in the linear regime is analogous to the thermal transport equation in the Fourier conduction limit, with the potential being analogous to temperature and the current being analogous the heat transfer rate.

For charge transport in CNTs in plastic, the substrate is considered insulating and only transport in the tube network is considered. For organic transistors with dispersed CNTs [40], the substrate is not insulating and charge leaks from the CNTs to the organic matrix, analogous to thermal transport in a composite, and charge exchange with the substrate must be considered. Since $L_C \gg \lambda$, the mean free path of electrons, a drift-diffusion model on Kirchoff's law for carrier transport may be employed [20]. In this linear regime, which occurs for low source-drain voltage V_{DS}, the current density along the tube is given by:

$$J = \sigma d\Phi/ds \tag{4}$$

where σ is the electrical conductivity and Φ is the potential, and is only a function of the source-drain voltage V_{DS}. Using the current continuity equation $dJ/ds = 0$ and accounting for charge transfer to intersecting tubes as well as to the substrate [41], the dimensionless potential distribution ϕ_i along tube i, as well the three-dimensional potential field in the substrate are given by:

$$\frac{d^2\phi_i}{ds^{*2}} + \sum_{\text{intersectingtubesj}} c_{ij}(\phi_j - \phi_i) + d_{is}(\phi_s - \phi_i) = 0, \tag{5a}$$

$$\nabla^{*2}\phi_s + \sum_{i=1}^{N_{tubes}} d_{is}\beta_v \frac{\sigma_t}{\sigma_s}(\phi_i - \phi_s) = 0. \tag{5b}$$

Here c_{ij} is the dimensionless charge-transfer coefficient between tubes i and j at their intersection point, analogous to Bi_c in Eq. (1a), and is specified a priori; it is non-zero only at the point of intersection. The term d_{is} is analogous to Bi_s term in Eq. (1a) and is active only for nanotubes in organic substrates. The electrical conductivity ratio is σ_t/σ_s. For computing the voltage distribution, boundary conditions $\phi_i = 1.0$ and $\phi_i = 0$ are applied to tube tips embedded in the source and drain regions respectively. For the organic substrate, $\phi_s = 1.0$ and $\phi_s = 0$ are applied at $x^* = 0$ and $x^* = L_C/d$ respectively; for the other boundaries, a treatment similar to that for the substrate temperature is applied. This computation of voltage distribution is only valid for low V_{DS}.

2.3 Solution Methodology

In the present analysis, the nanotube network is essentially 2D, while the substrate containing it is 3D, as shown in Fig. 2b. The source, drain and channel regions in Fig. 2b are divided into finite rectangular control volumes. A fixed probability p of a control volume originating a nanotube is chosen a priori. A random number is picked from a uniform distribution and compared with p. If it is less than p, a nanotube is originated from the control volume. The length of source and drain for tube generation is L_t, which ensures that any tube that can penetrate the channel region from either the left or the right is included in the simulations. The orientation of the tube is also chosen from a uniform random number generator. Since the tube length is fixed at L_t, all tubes may not span the channel region even for shorter channel lengths L_C, depending on orientation. Tubes crossing the $y^* = 0$ and $y^* = H/d$ boundaries are treated assuming translational periodicity; that part of the tube crossing one of these boundaries reappears on the other side. Tube-tube intersections are computed from this numerically generated random network and stored for future use. The analysis is conducted only on the tubes that lie in the channel region. The non-dimensional equations for the tubes and substrate are discretized using the finite volume method and a system of linearly coupled equations is obtained for the tube segment temperatures θ_i (or the electric potential ϕ_i) and the substrate temperatures θ_s (or the electric potential ϕ_s) at the substrate cell centroids. A direct sparse solver [42] is used to solve the resulting system of equations. To account for randomness in the sample, most of the results reported here are computed by taking an average over 100 random realizations of the network. More realizations are used for low densities and short channel lengths where statistical invariance is more difficult to obtain due to the small number of tubes in the domain.

3 Conduction in Percolating Network

If the underlying substrate (host matrix) has very low conductivity, the network itself forms the dominant pathway for conduction. This limit is realized in the case of electrical conduction in nanotube composites when the network is embedded inside an almost-insulating substrate such as plastic or glass. For thermal conduction in nanotube composites, the substrate-to-tube conductivity ratio is generally higher than for electrical conduction and substrate-to-tube interaction can be neglected only when the interfacial resistance between the tube and substrate is extremely high [19]. Therefore, the pure network conduction case is generally not realized for thermal transport.

The conductive properties of the network are strongly dependent on the density of the tubes in the network. A conducting path between source and drain may not exist at very low tube densities. If such tube network is used as the channel region of the transistor, no current could pass through the transistor. As the density of tubes increases, a critical density ρ_{th}, known as the percolation threshold, is reached, at which a complete pathway between source and drain is formed. The percolation threshold for the network is estimated as the density at which the average distance between the nanotubes equals the average length of the tubes, so that $\rho_{th} \propto 1 / \langle L_t \rangle^2$. A more accurate dependence of ρ_{th} on tube length, given by $\rho_{th} = 4.23^2/\pi L_t^2$, can be obtained from numerical simulations [43, 44]. There is great interest in exploring the transport behavior of the network at densities close to the threshold, which is dependent on the dimensionality and the aspect ratio of the tubes. Close to the percolation threshold, the network conductance, G, exhibits a power-law relation, i.e. $G \sim (\rho - \rho_{th})^m$, where m is the percolation exponent [45].

Several studies based on the Monte-Carlo simulations have been reported recently for the analysis of the percolating networks of nanotubes or their composites [29, 46]. Reference [29] analyzed the effect of contact resistance in percolation networks to explain why the value of the percolation threshold scaling exponent holds over the entire range of network-densities. Reference [46] performed Monte Carlo simulations to explore the aspect ratio dependence of the critical fractional volume and the critical index of conductivity. Reference [47] performed finite element analysis to explore the reasons for the absence of thermal percolation in nanotube composites. We have performed a percolation-based analysis to compute the conductance exponents of nanotube networks for different densities and for different tube-to-tube contact conductances [13]. We have also explored the change in the device performance when the semi-conducting tube network in the channel region is contaminated by metallic tubes [20]. Important results from these studies are summarized below.

3.1 Network Transport in the Non-Contacting Limit

In the limit when there is no contact between tubes ($Bi_c = 0$, $c_{ij} = 0$) and between tube and substrate ($\mathrm{Bi_s} = 0$, $\mathrm{d_{is}} = 0$) a simple analytical solution for the heat transfer rate through the domain (and correspondingly the drain current I_{DS} for electron transport) may be derived. Only the tubes are considered in this 2D planar calculation, and the substrate contribution is neglected. In this limit, the in-plane heat transfer rate through the composite, q, is directly proportional to the number of tubes directly bridging source and drain, but inversely proportional to the tube length contained in the channel. By computing the number of bridging tubes from geometric considerations, it may be shown that [13]:

$$q \propto I_{DS} \propto \frac{N_S}{W} = \left(\frac{2}{\pi}\right)\rho H L_t \left(\cos^{-1}\left(\frac{L_C}{L_t}\right) - \left(\frac{L_C}{L_t}\right)\sqrt{1 - \left(\frac{L_C}{L_t}\right)^2}\right). \tag{6}$$

where W is $\left[\sum_{i=1}^{N_S} {}^{1}/_{L_i}\right]^{-1}$, L_C is the channel length shown in Fig. 2, $\mathrm{L_t}$ is the length of the tube, and H is the height of the sample, as shown in Fig. 2, N_S is number of bridging tubes and ρ is the density of the tubes. The constant of proportionality in Eq. (6) depends on the conductivity of the tubes. Figure 3 shows a comparison of the analytical result obtained using Eq. (6) with that computed numerically using the finite volume method described above. The ratio q/q_{ref} is plotted, where q_{ref} is the reference heat transfer rate at $L_C/L_t = 0.1$. One hundred random realization of

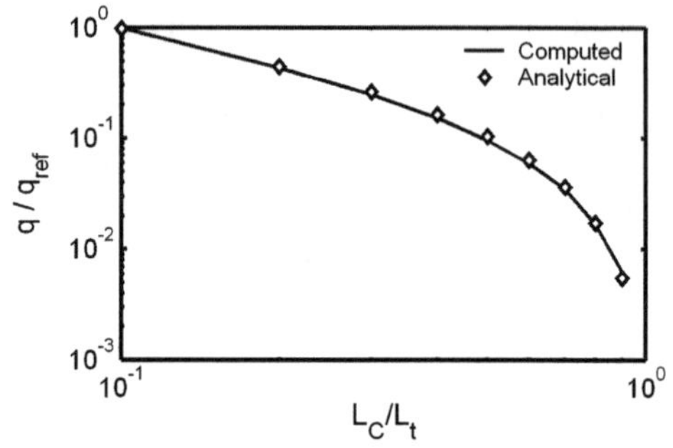

Fig. 3 Comparison of heat transfer rate in a nanotube network with analytical results for the case of zero tube-tube contact

the network are used. The case $L_C = 3$ μm, $H = 4$ μm, and $\rho = 5.0$ μm^{-2} is considered. The analytical and numerical results are in good agreement with each other, confirming the validity of our approach. When the channel length becomes comparable to or longer than the tube length, q/q_{ref} is seen to go to zero; in the absence of tube-tube and tube-substrate contact, heat or current can flow through the tubes only if the tubes bridge source and drain. As a practical matter, the result in Fig. 3 is applicable to electrical transport in short-channel CNT/plastic TFTs where the short channel lengths imply few tube-tube interactions.

3.2 Conduction Exponents

The lateral electrical conductivity of CNT thin films has been measured by different research groups. Here, we compare the network conductance predicted using our model with electrical conductance measurements by Ref. [11]. A pure planar tube network is considered, assuming that the substrate is entirely non-conducting. This is typical of electrical transport in CNT/plastic composites. The average length of the tubes in Ref. [11] ranges from 1 to 3 μm. The exact length distribution of nanotubes has not been reported in Ref. [11]. For the numerical model, random networks with a tube length of 2 μm are generated, and an average over 200 random realizations is taken. The percolation threshold for the network is roughly estimated using $\rho_{th} = 1 \big/ \langle L_t \rangle^2$ to be 0.25 μm^{-2}. Simulations are performed for densities in the range 1–10 μm^{-2} for channel lengths varying from 1 to 25 μm and with a width H of 90 μm, corresponding to the dimensionless parameters $L_C/L_t \sim 0.5$–12.5 and $H/L_t = 45$. The device dimensions and tube lengths are chosen to match those in Ref. [11].

In Fig. 4a, the normalized network conductance G/G_0 is shown as a function of L_C/L_t for several tube densities above the percolation threshold for nearly perfect tube-tube contact (i.e., $c_{ij} = 50$). For long channels $(L_C > L_t)$ there are no tubes directly bridging the source and drain, and current (heat) can flow only because of the presence of the network. If the tube density is greater than the percolation

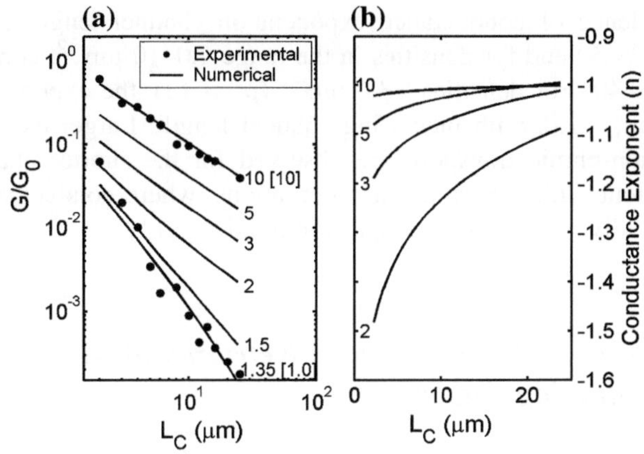

Fig. 4 **a** Computed conductance dependence on channel length for different densities (ρ) in the strong coupling limit ($c_{ij} = 50$) compared with experimental results from Ref. [11]. For $\rho = 10.0\ \mu m^{-2}$, $G_o = 1.0$ (simulation), $G_o = 1.0$ (experiment). For $\rho = 1.35\ \mu m^{-2}$, $G_o = 1.0$ (simulation), $G_o = 2.50$ (experiment). The number after each curve corresponds to the value of ρ used in the simulation. The number in [] corresponds to ρ in experiments from Ref. [11]. **b** Dependence of conductance exponent (n) on channel length for different densities (ρ) based on Fig. 4a

threshold, a continuous path for carrier transport exists from source to drain, and G is seen to be non-zero even for $L_C/L_t > 1$. Figure 4a shows that the conductance exponent, n, defined as $G \sim (L_C)^n$, is close to -1.0 for the high densities ($\rho = 10\ \mu m^{-2}$; $\rho^* = \rho/\rho_{th} = 40$), indicating ohmic conduction, in good agreement with Ref. [11]. The exponent increases to -1.80 at lower densities ($1.35\ \mu m^{-2}$; $\rho^* = 5$), indicating a non-linear dependence of conductance on channel length. The asymptotic limit of the conductance exponent for infinite samples with perfect tube/tube contact has been found to be -1.97 [48, 49]. The observed non-linear behavior for low density is expected because the density value is close to the percolation threshold. Snow et al. reported a conductance-exponent of -1.80 for a density of $1.0\ \mu m^{-2}$ and channel length >5 μm. For the same device dimensions, this value of the exponent is close to that obtained from our simulations for a density of $1.35\ \mu m^{-2}$. At densities close to the percolation threshold, computations are very sensitive to variations in computational parameters. Small variations in experimental parameters such as tube diameter, nanotube contact strength, tube electronic properties as well as the presence of a distribution of tube lengths (1–3 μm), which is not included in the simulation, may explain the difference. The contact resistance between the nanotubes and the source and drain electrodes as well as insufficiently large samples for ensemble averaging in the experimental setup may also be responsible. Indeed, a more quantitative agreement with the low-density data at short channel length is realized if one accounts for imperfect tube-tube contacts [50].

The dependence of conductance exponent on channel length is explored in Fig. 4b for $c_{ij} = 50$ and for densities in the range 2.0–10 μm^{-2}, corresponding to ρ^* values of 8–40. For densities >3.0 μm^{-2} ($\rho^* > 12$), the exponent approaches the ohmic limit, -1.0, with increasing channel length. Larger exponents, corresponding to non-ohmic transport, are observed for the shorter channel lengths. This is consistent with experimental observations, where conductance is seen to scale more rapidly with channel length for small L_C [11].

3.3 Conduction in Heterogeneous Networks of Metal-Semiconducting Tubes

The electrical performance of CNT networks is strongly influenced by the fact that approximately one-third of the CNTs grown by typical processing techniques exhibit metallic behavior and approximately two-third exhibit semiconducting behavior [20]. This heterogeneity controls the on-off ratio R of typical CNT-network based devices, where on-off ratio R is the ratio of the device-current in the on state to the device current in the off-state. R has been shown [20] to be a unique and predictable function of L_C, L_t, N_{IT} (the density of interface traps), f_M, (the degree of metallic contamination) and ρ, the tube density. In the conventional transistors, N_{IT} is the trapped charge at the interface of the channel and the insulating dielectric SiO_2, which separates the gate from the channel. f_M is the ratio of the number of metallic tubes to semi-conducting tubes in the tube-network. If the on-off ratio R can be reliably predicted as a function of tube density ρ and other parameters, our numerical model affords a unique way to find the tube density of typical CNT thin films by using this relationship in the inverse. This method promises far more accurate estimation of 'electrically relevant' tube density than methods currently in use, such as atomic force microscopy, and scanning electron microscopy [11].

We compute I_{DS} versus V_{GS} for several tube densities ($\rho = 1$–5 μm^{-2}) for device parameters $L_C = 10$ μm, $L = 2$ μm, $H = 35$ μm, and $V_{D\ S} = 0.1$ V (Fig. 5) which are chosen to match the experiments in [11]. Since $L_C >> L_t$, these transistors are called long-channel devices (note that this terminology differs from classical transistor terminology—where long and short channels are defined with respect to electrostatic control of the channel by the gate electrode [51]. We use $c_{ij} \sim 50$ based on typical values for CNT tube-tube contact [11, 52], mobility [11] and density functional theory. Here, f_M, is taken to be 33 % in Fig. 5, consistent with Ref. [11]. The conductance ratio of metallic to semiconducting tubes (M/S conductance ratio) in the on-state is chosen as 8.0, consistent with Ref. [53]. In general, the M/S conductance ratio depends weakly on the fabrication process, as well as the chirality, band-gap and the diameter of the tubes.

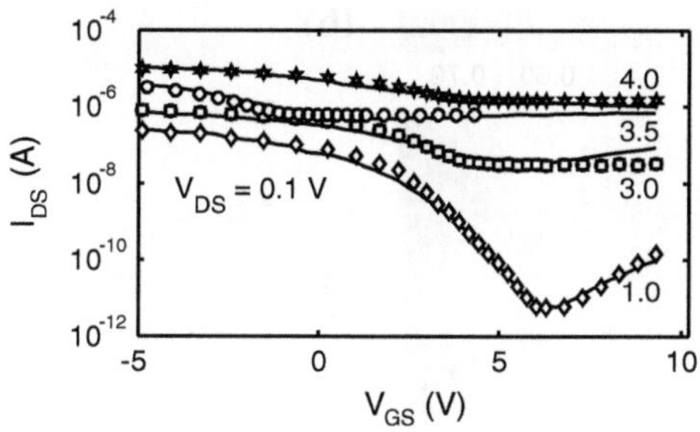

Fig. 5 Computed $I_{DS} \sim V_{GS}$ at $V_{DS} =$ 0.1 V for different densities is compared with experimental results from Ref. [11] before the electrical breakdown of metallic tubes. *Solid lines* correspond to experimental results from Ref. [11] and markers correspond to computational results. The number after each curve corresponds to tube density ρ. The curve $\rho = 3.5\ \mu m^{-2}$ is shifted on the x-axis to account for charge trapping

Gate characteristics, represented by I_{DS}–V_{GS} curves, are computed for a specific network configuration. An average is then taken over 50 random realizations of the network. Computations for $\rho = 1\ \mu m^{-2}$ agree very well with experiments in Ref. [11], Fig. 5. Increasing ρ increases the number of percolating metallic paths, increasing the on-current I_{ON}, but reducing *R*, as in Ref. [11]. Snow et al. speculate that $\rho > 3\ \mu m^{-2}$ for devices with low on-off ratio (top three solid lines in Fig. 5). Our simulations establish that they correspond to exact densities of $\rho = 3.0$, 3.5 and 4.0 respectively. Thus, tube density ρ may be deduced from a simple electrical measurement of the on/off current ratio (see Fig. 5) obviating the need for inaccurate and time-consuming analysis of AFM images, as is currently done. We note, however, that although we can predict $R(\rho)$ for a fixed M/S conductance ratio and c_{ij}, the absolute value of the on-off current and *R* can still vary from sample to sample depending on the M/S conductance ratio and the contact conductance between tubes of different diameters. The same methodology can also be used to interpret short channel data (Fig. 2 in Ref. [8], for example). This demonstrates the predictive power of the theoretical framework.

4 Conduction in Nanotube Composites

Thus far, we have considered conduction in a pure network of nanotubes in the absence of a substrate or host matrix. We now turn our attention to thermal and electrical conduction in composites where carrier transport is no longer confined exclusively to nanotube network. When the thermal conductivity ratio k_s/k_t is significant or when the heat leakage from the tubes to the substrate is significant, both substrate and network play important roles in determining the thermal performance of typical nanotube composites. For electrical transport, the substrate does not generally play a role in nanotube-polymer composites, since the polymer

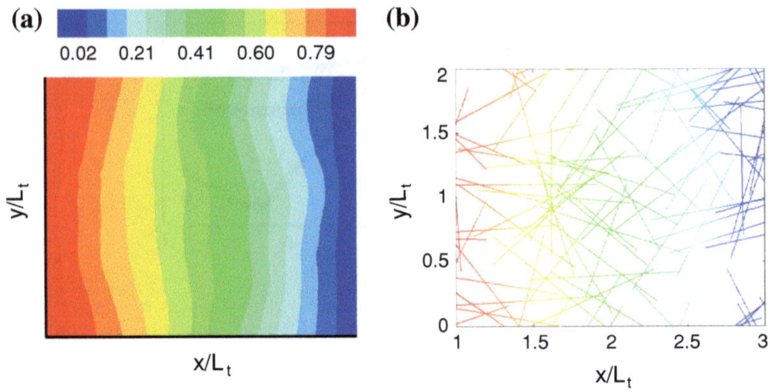

Fig. 6 Non-dimensional temperature distribution in (**a**) substrate (**b**) tube network. $L_C/L_t = 2.0$, $H/L_t = 2$, $Bi_c = 10.0$, $Bi_s = 10^{-5}$, $k_s/k_t = 0.001$ and $\rho^* = 14.0$

is essentially insulating. However, a problem analogous to thermal transport occurs in electrical transport in nanotube-organic composites. Here, the electrical conductivity of the organic substrate is relatively large, and forms the primary conduction pathway. Recently, sub-percolating nanotube dispersions have been added to enhance electrical conductivity of organics [40].

Using the formulation described in Sect. 3, a typical temperature distribution in the tube network and the substrate is computed and shown in Fig. 6a, b. For this case, $L_C/L_t = 2.0$, $H/L_t = 2$, $Bi_c = 10.0$, $Bi_s = 10^{-5}$, $k_s/k_t = 0.001$ and $\rho^* = 14.0$, corresponding to $L_C = 4$ μm, $L_t = 2$ μm, $H = 4$ μm, and $\rho = 3.5$ μm^{-2}. Contours of constant temperature in the substrate would be one-dimensional in x for $Bi_s = 0$, but due to the interaction with the tubes, distortion in the contours is observed, consistent with the temperature plots in the tube in Fig. 6b. The departure from one-dimensionality in the substrate temperature profile is related to local variations in tube density; regions of high tube density convey the boundary temperature further into the interior.

4.1 Effect of Tube-Substrate Interfacial Resistance

The contact parameters Bi_S for tube-substrate contact and Bi_C for tube-tube contact are difficult to determine, and there are few guidelines in the literature to choose them. Recent experimental studies conducted in Ref. [54] suggest that heat transport in nanotube composites may be limited by exceptionally small interfacial thermal conductance values. The value of the interfacial resistance between the carbon nanotube and the substrate was reported to be 8.3×10^{-8} m^2 K/W in Ref. [54] for carbon nanotubes in hydrocarbon (do-decyl sulphate). The non-dimensional contact parameter Bi_S evaluated using this contact conductance is O(10^{-5}) assuming the

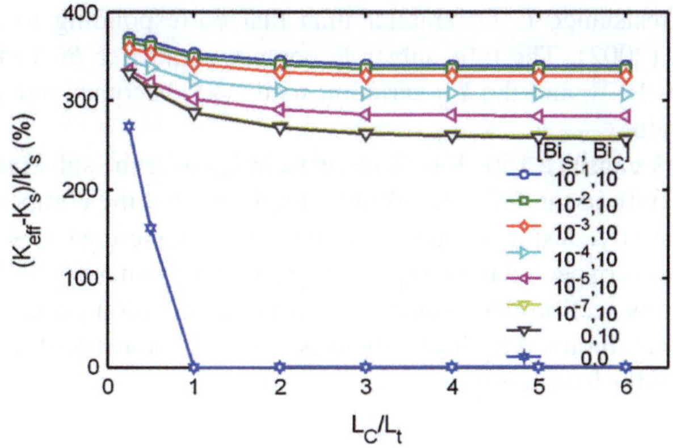

Fig. 7 Effect of substrate-tube contact conductance (Bi_S) on k_{eff} for varying channel length. $L_C/L_t = 0.25$–6.0 ($L_t = 2.0$ μm), $H/L_t = 2$, $k_S/k_t = 0.001$, and $\rho^* = 3.5$ ($\rho = 5.0\ \mu m^{-2}$)

thermal conductivity of a single SWCNT to be 3,000 W/mK. The corresponding value for nanotube-polymer or nanotube-glass interfaces is not known at present. Consequently, values of Bi_S in the range 10^{-1} to 10^{-7} were considered. If we assume the thermal conductivity of the polymer matrix to be 0.25 W/mK, a value of $Bi_S = 10^{-5}$ corresponds to the thermal resistance of an equivalent polymer layer of thickness 20 nm.

The percentage increase in k_{eff} of the composite is plotted against L_C/L_t ratio for different Bi_S in Fig. 7 for, $L_C/L_t \sim 0.25$–6.0 ($L_t = 2.0$ μm), $H/L_t = 2$, $k_S/k_t = 0.001$, and $\rho^* = 3.5$ ($\rho = 5.0\ \mu m^{-2}$). Here, the tube density is much higher than the percolation threshold density, $\rho_{th} \sim 1.42\ \mu m^{-2}$. The tube-tube contact parameter is held at $Bi_C = 10$, denoting nearly perfect contact. The conductivity ratio is 10^{-3}, denoting highly conducting tubes in a relatively insulating substrate. A sharp increase in k_{eff} is observed for shorter channel lengths as a result of highly conducting tubes directly bridging source and drain. This increase for shorter channel lengths is less significant at high Bi_S due to high heat leakage from the tubes to the substrate. As the channel length increases, the composite approaches bulk behavior and the conductance achieves invariance beyond $L_C/L_t > 5$. The asymptotic values of k_{eff} at high channel lengths ($L_C/L_t > 5$) can be expressed as $k_{eff} \sim 1 + \gamma k_S/k_t$, where γ is dependent on Bi_S, Bi_C and area-ratio of tubes cross-section and substrate at a composite cross-section. The general shape of the curves is explained by the 'zero Bi_S, zero Bi_C' case. As the channel becomes narrower, bridging occurs and so the curve rises. This is a "finite-size" effect not seen in bulk composites, and exists even for channels 3–4 times the tube length.

It is observed that the curve for $Bi_S \sim 10^{-7}$ is the same as that for $Bi_S \sim 0$, Fig. 7. This defines the lower limits of tube-substrate contact—below this only the network is active and side-leakage disappears. For this case, the only reason for the existence of a non-zero k_{eff} increase is the network. In this limit we would expect to see percolation behavior, unlike that noted in the literature [25]. This implies that the interface resistance is far smaller than that corresponding to $Bi_S = 10^{-7}$ in Biercuk et al. (2002). The tube-substrate contact parameter Bi_s ceases to be limiting for $Bi_S > 10^{-2}$, and the k_{eff} variation with L_C/L_t becomes independent of Bi_S beyond this value.

The network causes a 150–350 % increase in k_{eff} over the substrate, but this still means values only in the 0.35–3.5 W/mK, implying that the composite is not very conducting laterally, despite the presence of highly conducting tubes. Since the k_S/k_t value used here is expected to be typical of many composites for TFT applications, the results in this section demonstrate that if the percolation properties of the network could be maintained by high tube-tube contact, the network itself could provide a pathway for heat removal.

4.2 Effect of Tube-Tube Conductance

Recently, Lukes and co-workers [36] considered heat transfer between CNTs using classical molecular dynamics simulations, and estimated a tube-tube contact resistance of the order of 1.0×10^{-7} m^2-K/W, corresponding to a Bi_C value of 2.0×10^{-5} (assuming $k_t = 3{,}000$ W/mK). However, there is no experimental corroboration of tube-tube contact resistance, and Bi_C values of 0–10^{-1} are chosen, ranging from insulating contact to perfect contact. Previous theoretical models for computing k_{eff} of nanotube composites ignore contact conductance between the nanotubes [24]. Since tube-tube contact occurs over a very small contact area, this contact resistance would be limiting only in the case of large tube-substrate resistance ($Bi_S \rightarrow 0$) for most k_S/k_t values of interest. Figure 8 show the $k_{eff} \sim L_C/L_t$ plots for different Bi_C for $L_C/L_t \sim 0.25$–6.0 ($L_t = 2.0$ μm), $H/L_t = 2.0$, $k_S/k_t = 0.001$, $Bi_S = 10^{-5}$, and $\rho^* = 3.5$ ($\rho = 5.0$ μm^{-2}). The overall shape of the curves is similar to that in Fig. 7. The sharp increase in k_{eff} for shorter channel length composites (due to bridging tubes) is significant for low Bi_C, but the effect gradually diminishes with increasing Bi_C, Fig. 8. Decreasing Bi_C from 10^{-1} to 10^{-5} decreases k_{eff} by about 50 % for long channels. Beyond $Bi_C > 10^{-2}$, tube-tube contact is sufficiently good that it ceases to matter; consequently the k_{eff} curves become coincident in Fig. 8. By the same token, values of Bi_C below 10^{-5} mean essentially zero contact between tubes, and heat leakage through the substrate is the only pathway for heat transfer for long channels. The tube-tube contact resistance computations by Zhong and Lukes [36] suggest that this may indeed be the mechanism of heat transfer in CNT composites, though further experimental corroboration is necessary. The present analysis reveals that for tube densities higher than the percolation threshold, Bi_C may be an important parameter controlling k_{eff} for highly conducting tubes with high tube-substrate resistance. Bi_C (10^{-2}–10^{-5}) corresponds to the thermal resistance presented by an equivalent polymer matrix of thickness 0.025–25 nm assuming contact area between the tubes of the order $\sim d^2$ and $k_S/k_t = 10^{-3}$.

Fig. 8 Effect of tube-tube contact conductance (Bi_C) on k_{eff} for varying channel length. $L_C/L_t = 0.25$–6.0 ($L_t = 2.0$ μm), $H/L_t = 2$, $k_S/k_t = 0.001$, $Bi_S = 10^{-5}$, and $\rho^* = 3.5$ ($\rho = 5.0$ μm^{-2})

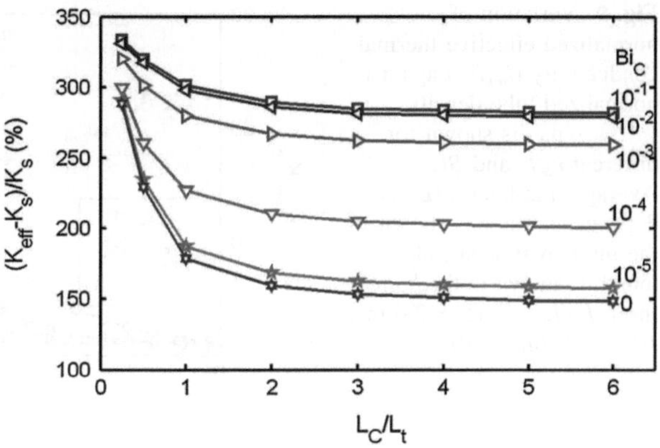

4.3 *Effect of Tube Density*

The effective thermal conductivity of the composite is observed to have a different dependence on ρ in different regimes, i.e. $\rho \ll \rho_{th}$, $\rho \sim \rho_{th}$ and $\rho \gg \rho_{th}$ [55]. k_{eff} is shown as a function of ρ for these three different regimes in Fig. 9 for the case ρ_{th} ($= 4.23^2/\pi L_t^2 \sim 1.4$ μm^{-2}), $k_S/k_t \sim 10^{-4} - 10^{-1}$, $Bi_S \sim 10^{-7} - 10^{-2}$, $L_C/L_t = 3\,(L_t = 2.0\,\mu\text{m}), H/L_t = 2, Bi_C = 10$. When $\rho \ll \rho_{th}$, k_{eff} increases linearly with ρ. This is expected since the tubes do not interact with each other either through direct contact or through the substrate. This lack of tube-tube interaction is borne out by the variation of the average *link length* (L_{link}) with ρ^* (inset in Fig. 9), which shows that the non-interacting limit between tubes is achieved for $\rho^* \ll 1$ for which $L_{link}/L_t \sim 1$. This trend is in agreement with the results from EMA which predict a linear scaling with volume fraction. When $\rho \sim \rho_{th}$, k_{eff} is observed to vary non-linearly with ρ. This is typical of network percolation close to the percolation threshold [11, 13] and becomes more pronounced with decreasing Bi_S *or* k_S/k_t, Fig. 9. These two parameters explain the difference between percolation behavior for thermal transport and for electrical transport. Strong non-linear behavior near the percolation threshold is observed for charge transport in CNT-polymer composites due to very low k_S/k_t ($<10^{-6}$) [11, 13], while for thermal transport, this non-linear behavior is relatively weak due to high k_S/k_t ($\sim 10^{-3}$) and high heat leakage through the substrate (high Bi_S). Whenever transport through the substrate competes with the transport through the CNT network either due to high k_S or due to high heat leakage from the CNTs to the substrate, percolation effects due to the network are suppressed.

For $\rho > 3.0\,\rho_{th}$ and for large enough L_C/L_t, k_{eff} is found to vary linearly with ρ, Fig. 9. The reason for this is evident in the inset in Fig. 9, which shows that the average L_{link} varies linearly with ρ^* for high densities ($\rho^* > 3$). Hence the network becomes homogenous. Computations of k_{eff} for the pure network in the absence of

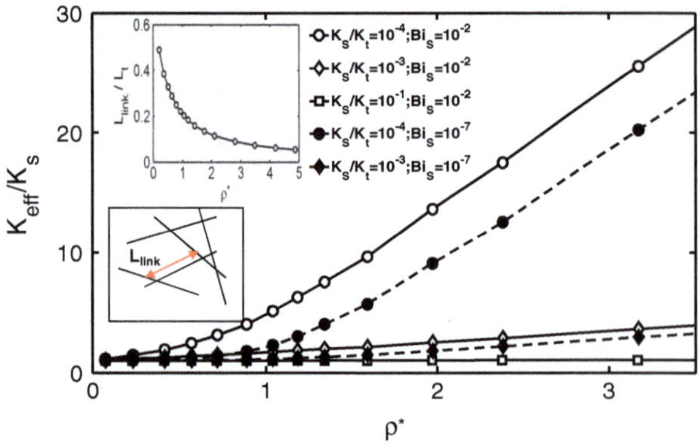

Fig. 9 Variation of normalized effective thermal conductivity (k_{eff}/k_S) against normalized tube density ($\rho^* = \rho/\rho_{th}$) is shown for different k_S/k_t and Bi_S . Average link-length (L_{link}) dependence on ρ is shown in the inset. A typical link length is shown in the *bottom* inset. $L_C/L_t = 3$ (L_t= 2.0 m), $H/L_t = 2$, $Bi_C = 10$

the substrate [13] reveal that it may be expressed as $k_{eff}/k_t \sim \rho L_t^2(0.783 - 0.119 \ln(Bi_C^2) - 0.015 \ln(Bi_C))$ for high densities. For the CNT composites, this expression would also depend on Bi_S and k_S/k_t, but linearity with respect to density would nevertheless be valid.

4.4 Electrical Conductivity of CNT-Organic Composites

A novel approach involving modifying the transconductance ($g_m \sim dI_{DS}/dV_{GS}$) of an organic host using a sub-percolating dispersion of CNTs has been proposed in Ref. [40]. A 60-fold decrease in effective channel length, L_{eff}, is observed that results in a similar increase in g_m with a negligible change in on-off ratio [40]. In this technique, the majority of the current paths are formed by the network of CNTs, but short switchable semiconducting links are required to complete the channel path from source to drain [40]. Experimental data published in Ref. [40] provide a good opportunity to test the correctness of our numerical formulation in Sect. 3.

The device parameters $L_C = 20$ μm, $L_t = 1$ μm, and $V_{DS} = -10$ V are chosen to match the experiments in Ref. [40]. Charge transfer coefficients $c_{ij} = 10^{-4}$ and $d_{is} = 10^{-4}$ are assumed and correspond to poor contact conductance between tube-tube and tube-substrate. The electrical conductivity ratio, σ_t/σ_s, for metallic CNTs in the on-state ($V_{GS} = -100$ V) is taken as 5.0×10^4, while that for semiconducting CNTs is 5.0×10^3 [53]. The metallic-CNT conductivity is assumed constant with V_{GS}, while the roll-off in the conductivity of semiconducting CNTs and the organic- matrix with V_{GS} is obtained from the experimental I_{DS}–V_{GS} curves (0 % and 0.5 % volume fraction curves) in Fig. 1b of Ref. [40]. Figure 10 shows that numerical results agree well with experiments over the entire range of tube densities (1.5–17 μm^{-2}). There is an anomalous jump in the I_{DS}–V_{GS} curve for 0.5 % volume fraction of CNTs (labeled "shift" in Fig. 10; see also

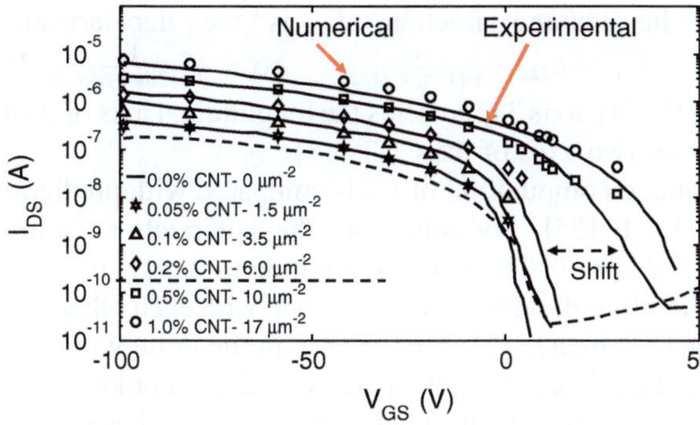

Fig. 10 Computed I_{DS}–V_{GS} at $V_{DS} = -10$ V for different CNT-densities ($\rho \sim 1$–$17\ \mu m^{-2}$) is compared with experimental results in [40]. The volume % of CNT dispersions used in the experiments and the corresponding network density (ρ (μm^{-2})) used in the computations are shown. $L_t = 1\ \mu m$, $L_C = 20\ \mu m$ and $H = 200\ \mu m$. The shift in the I_{DS}–V_{GS} curves due to the initiation of semiconducting CNT percolation for CNT volume % > 0.2 % is shown by the *dashed arrow*

Fig. 1b in Ref. [40]) which is not properly understood. We calculated the I_{DS}–V_{GS} characteristics of the organic TFT device in [40] with a realistic heterogeneous network of semiconducting-metallic tubes (1:2 ratio) dispersed in an organic matrix. We have shown that this anomalous shift in the I_{DS}–V_{GS} curve is a consequence of the formation of a parallel sub-percolating network of *semiconducting* CNTs in the organic matrix. At 0.2 % CNT volume fraction, the semiconducting tubes do not have sufficient density to form a percolating network in and of themselves; metallic CNTs are necessary to achieve percolation. However, when the volume fraction is increased to 0.5 %, semiconducting tubes can form a percolating network by themselves, and shift the I_{DS}–V_{GS} curve as shown. This confirms that semiconducting CNTs are active elements of this organic TFT device, a feature which was not understood previously.

4.5 Comparison with Effective Medium Theory

For low-density dispersions, the effective conductivity (k_{eff}) can be derived using a Maxwell-Garnett effective medium approximation [18], and provides a baseline for comparison with our numerical calculations. For a planar CNT network isotropic in the x-y plane (see inset in Fig. 11b) and embedded in a substrate of thickness *d,* the theory in [18] may be modified to yield k_{eff} in the x-y plane as

$$k_x = k_y = k_S \frac{2 + f[\beta_{11}(1 - L_{11}) + \beta_{33}(1 - L_{33})]}{2 - f[\beta_{11}L_{11} + \beta_{33}L_{33}]}. \tag{7}$$

Here f is the volume fraction, L_{ii} is the depolarization factor, and $\beta_{ii} = \frac{k_{ii}-k_S}{k_S+(k_{ii}-k_S)}$; $k_{11} = k_{22} = \frac{k_t}{1+(2a_K k_t/k_S d)}$; $k_{33} = \frac{k_t}{1+(2a_K k_t/k_S L_t)}$. Here, a_K is the Kapitza radius [18, 24], axis 3 represents the longitudinal axis of the CNT and axes 1 and 2 are the other two axes of the CNT [18].

The finite volume computation of k_{eff} is compared with predictions from the 2D EMA in Fig. 11a, b [55]. For this case, the polarization factors are given by $L_{11} = L_{22} = 0.5, L_{33} = 0$. The basic assumption in Eq. (7) is that tube density ρ is very low, and therefore the tubes do not interact with each other. Consequently, the tube-tube contact parameter Bi_C is set to zero in the finite volume computations to obtain a direct comparison. Since the parameter Bi_S is not known a priori, its value is adjusted to match the results from EMA for a value of $a_K = 0.25$ nm for $k_S/k_t = 10^{-2}$. The same value of Bi_S is used in all subsequent calculations for other k_S/k_t ratios in Fig. 11a. It is important to notice that there is only one free parameter, Bi_S, in our simulations, corresponding to the adjustable parameter a_K in EMA. A good match with the results of EMA is obtained for the case of $Bi_C = 0$. Calculations were also performed in Fig. 11a for $Bi_C = 10$, representing nearly-perfect contact. For high Bi_C, the numerically computed k_{eff} is observed to deviate substantially from the EMA prediction even for densities below the percolation threshold ρ_{th}. This deviation is significant for all but the highest k_S/k_t values ($<10^{-2}$), and would therefore be significant for computations of electrical and thermal conductivities in CNT composites. This suggests that high aspect ratio tubes strongly interact with each other even at tube densities below ρ_{th} and that the EMA is inadequate for the prediction of k_{eff} at all but the very lowest densities. Figure 11b presents the variation of k_{eff} with tube density with Bi_S as a parameter; the ratio $(1/a_K)/Bi_S$ is held constant in the figure. A good match with EMA is found. The constant ratio $(1/a_K)/Bi_S$ for different curves in Fig. 11b shows that the adjustable parameter in EMA $(1/a_K)$, and that in the numerical calculations, Bi_S, are consistent. These results indicate that adjusting a_K to fit experimental data for $\rho > \rho_{th}$ in previous studies [24, 56] adjusts in part for tube-tube interaction effects not present in EMA theory. These adjustments would tend to underpredict the true value of interface resistance, a claim also supported by Ref. [57].

5 Interfacial Thermal Transport Between Nanotubes

In the last two sections, we have considered conduction in a network of nanotubes and their composites where we have treated tube-tube and tube-substrate resistance values as parameters. Various experimental and numerical studies have been performed to analyze the interfacial energy transport and to estimate the thermal resistance between the CNTs and between the CNT and the substrate. In this section, we present our recent investigation of the thermal energy transport from one single-walled nanotube (SWNT) to another SWNT positioned in a crossed configuration and subjected to an intense heat pulse using Molecular Dynamics

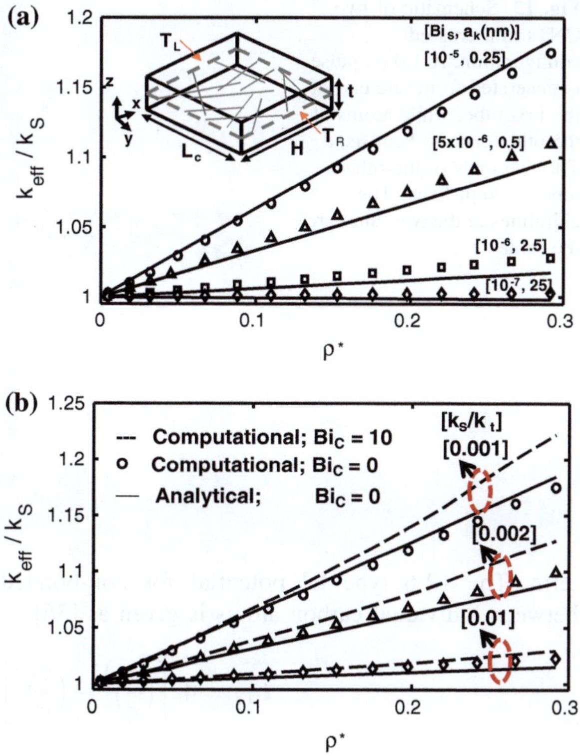

Fig. 11 **a** Comparison of normalized effective thermal conductivity (k_{eff}/k_S) computed from numerical simulations (markers) and analytically-derived expressions (*solid lines*) for different k_S/k_t ratios. $L_C/L_t = 8$ ($L_t = 0.5$ μm), $H/L_t = 4$, $Bi_S = 10^{-5}$ and $\rho^* = 0–0.3$ ($\rho = 0.1–6.5$ μm^{-2}). $Bi_C = 0$ for simulations unless otherwise stated. **b** Comparison of normalized effective thermal conductivity (k_{eff}/k_S) computed from numerical simulations and 2D EMA for different values of the interfacial thermal resistance Bi_S (10^{-7}–10^{-5}). $L_C/L_t = 8$ ($L_t = 0.5$ μm), $H/L_t = 4$, $k_S/k_t = 10^{-3}$ and $\rho^* = 0–0.3$ ($\rho = 0.1–6.5$ μm^{-2})

(MD) techniques [58]. A schematic of two CNTs in a crossed configuration is shown in Fig. 12. Here, the CNTs are placed perpendicular to each other with a gap equal to a van der Waals distance of 3.4 Å. These CNTs are not covalently bonded at the interface.

Selection of appropriate interatomic energies and forces is important for the reliability of classical MD simulations. We use the reactive empirical bond order (REBO) potential for C-C bond interaction and a truncated 12-6 type Lennard-Jones potential for non-bonded van der Waals interactions between CNTs. The REBO potential has been extensively applied to perform MD simulations in CNTs and CNTs in hydrocarbon composites/suspensions [31]. The analytical form of this potential is based on the intramolecular potential energy originally derived by Abell [59]. The REBO potential is given by:

$$U_{REBO} = \sum_i \sum_{i<j} [V_r(r_{ij}) - D_{ij} V_a(r_{ij})]. \quad (8)$$

where r_{ij} denotes the distance between atoms i and j, V_r corresponds to interatomic core-core repulsive interactions, and V_a describes the attractive interactions due to the valence electrons. Here, D_{ij} corresponds to a many-body empirical bond-order

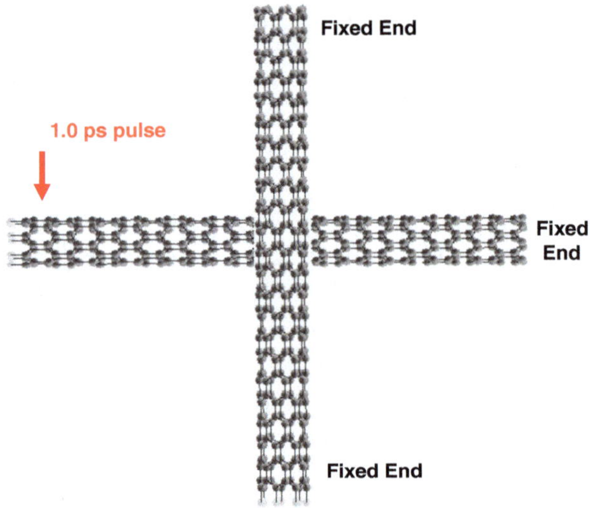

Fig. 12 Schematic of two CNTs in a crossed configuration. A 1.0 ps pulse is generated from one end of the first tube, while atoms at the other end are kept fixed. The two ends of the other tube are kept fixed. The chiralities of the two tubes are the same

term. The 12-6 type LJ potential for non-bonded van der Waals interaction between individual carbon atoms is given as [36]:

$$U_{LJ} = 4\varepsilon\left[\left(\frac{\sigma}{r}\right)^{12} - \left(\frac{\sigma}{r}\right)^{6}\right]. \tag{9}$$

Several different values of the energy and distance parameters in the L-J potential are considered for the interaction of C-C atoms in the CNT. The present study employs the parameterization used by Lukes et al., with $\varepsilon = 4.41$ meV and $\sigma = 0.228$ nm. The details of the MD code used for the present analysis may be found in Ref. [59].

5.1 Heat Pulse Analysis Using Molecular Dynamics

MD simulations are used to examine transient heat pulse propagation in zig-zag tubes positioned in a crossed configuration for chiralities varying from (5,0) to (10,0). We generate the heat pulse at one end of the tube using the methodology proposed by Osman and Srivastava [60] for studying energy transport through a single CNT at low temperatures. In order to study heat pulse propagation in CNTs and to compare the results for different chiralities, each CNT is divided into 500 slabs along its axis (each slab is a ring). The length of each CNT is 106 nm, but the number of atoms in a slab depend on the chirality or diameter of the CNT, i.e., a single slab in a (5,0), (6,0), (7,0), (8,0) and (10,0) CNT would have 10,12,14,16, and 20 atoms respectively. One end of the first CNT, where the heat pulse is generated, is treated as a free boundary, while the other end is kept rigid (see Fig. 12); both ends of the second CNT are held rigid. The boundary region of the first CNT (i.e. the CNT in which the pulse is generated) extends over 10 slabs. Before the generation of the heat pulse, both CNTs are quenched to a very low temperature of 0.01 K for 50,000 time steps (25 ps) to achieve thermal equilibrium at 0.01 K. Then, a strong heat pulse of one picosecond duration and a peak temperature of 800 K is generated at one end of the first CNT. The heat pulse is applied to 10 slabs near the left boundary using a Berendsen thermostat [61]. The heat pulse consists of a 0.05 ps rise time, a 0.9 ps duration with a constant temperature of 800 K, and a 0.05 ps fall time. During the 0.05 ps fall time, the temperature of the boundary slabs is decreased to reach a final temperature of 0.01 K and then held constant at that temperature for the rest of the simulation. This is done to prevent the exchange of large amounts of energy from the boundary slabs to the region of interest after the generated pulse has started propagating towards the right boundary of the first CNT.

Our analysis is focused on the time window from the point of generation of the pulse to the time before heat pulse reaches the right boundary to avoid the effects of boundary reflection. The temperature of each slab is spatially averaged over ten slabs centered at the slab of interest. In our simulations, the temperature is time-averaged over sets of 200 time steps ($\sim$100 fs) to reduce the effect of statistical fluctuations and is recorded during the entire simulation time. The speed of the pulse is determined from the spatial distance traversed by the particular pulse during a given time interval.

We first study the interaction between two CNTs in a crossed configuration when a heat pulse is passed through first CNT using the methodology described in the above sections. Our interest is in analyzing the energy transfer to the second tube when the heat pulse passes through the contact zone, and also to study the waves generated in the second tube due to this energy transfer. The distance between the CNTs remains in the range of 3–3.7 Å during the entire simulation for CNTs of chirality (5,0), (6,0) and (7,0). For CNTs of chirality (8,0), this distance remains in the range of 4–4.5 Å, while for CNTs of chirality (10,0), it remains in the range of 5–5.5 Å.

The location and shape of the heat pulses at different time instants along the first and second CNTs for the case of (5,0) chirality are shown in Fig. 13a, b respectively. The amplitude of the heat pulses is presented in terms of the average kinetic energy of the atoms in ten slabs at any location. Here, $t = 0$ ps corresponds to the time when heat pulse generation has started at the left end of the first CNT. The attenuation in peak kinetic energy ($\sim$1.2 eV) of the heat pulse is negligible and it is seen to propagate like a ballistic pulse along the first tube. This pulse moves with a speed of 22 km/s along the nanotube, which is very close to the speed of sound (20.3 km/s) associated with longitudinal acoustic (LA) phonon waves in zig-zag nanotubes [60]. Two CNTs located in a crossed configuration make contact at their mid-point, which is at 53 nm from their free ends. When the heat pulse in the first tube approaches this contact zone, the kinetic energy in the other tube increases at the location of the contact, Fig. 13b. The kinetic energy reaches its maximum at $t = 2.6$ ps which is approximately the time taken by the heat pulse in the first tube

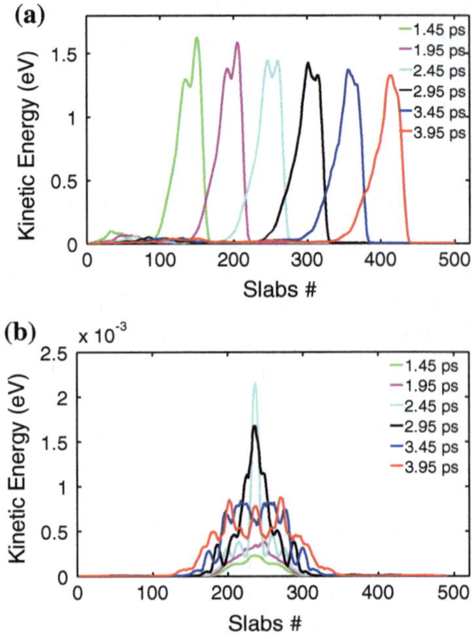

Fig. 13 **a** Location and shape of the heat pulse at different time instants along the first CNT for (5,0) chirality. The pulse generation starts at t = 0.0 ps. **b** Location and shape of the heat pulse at different time instants in the second CNT, also of the same chirality

to cross the contact zone. The peak kinetic energy in the second tube is 1.5 meV, which is very low in comparison to peak kinetic energy of 1.2 eV of the heat pulse in the first tube. This indicates that coupling is very weak between two CNTs for these fast moving pulses, since little time is spent by the pulse in the contact zone. The energy given to the second tube at the point of contact spreads along the second tube. This is indicated by the decreasing kinetic energy at the center of the second tube and the symmetrically excited heat pulses on both sides of the contact in Fig. 13b. The speed of propagation of the excited heat pulse in the second tube is very low and it is difficult to relate these pulses with any specific phonon mode by observing their transient propagation profiles.

We perform a similar heat pulse analysis for nanotubes of chirality (6,0), (7,0), (8,0) and (10,0); the shape and location of the heat pulse along the nanotube for different time instants for the (7,0), (8,0) and (10,0) chiralities are shown in Figs. 14, 15 and 16. It is observed that the behavior of heat pulses in the nanotubes is very dependent on the chirality or the diameter of the tube. The speed of heat pulse propagation in the first tube and the increase in kinetic energy in the second tube until t = 5.5 ps are listed in Table 2. Heat pulse propagation in the first tube for (6,0) and (7,0) chiralities is similar to that for the (5,0) nanotube, i.e., the heat pulse propagates like a ballistic wave with a speed in the range of 22–23 km/s. However, the pulses excited in the second tube are significantly different for different chirality tubes. No conclusions can be drawn merely by looking at the kinetic energy rise in the second tube (see Table 2). However, it can be observed that heat spreading along the second tube is faster in the (5,0) tube in comparison

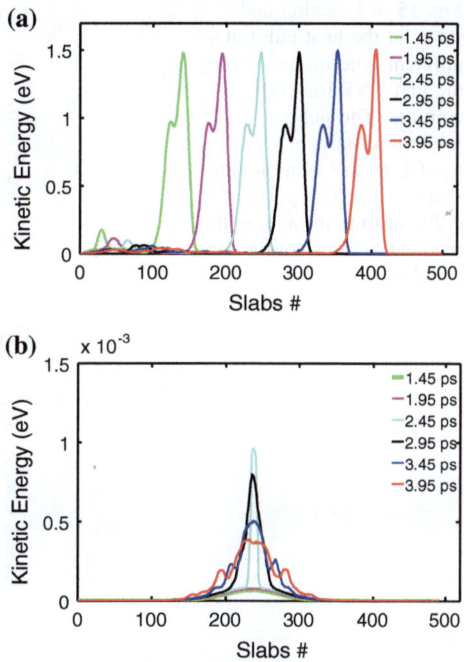

Fig. 14 **a** Location and shape of the heat pulse at different time instants along the first CNT for (7,0) chirality. The pulse generation starts at t = 0.0 ps. **b** Location and shape of the heat pulse at different time instants in the second CNT, also of the same chirality

to the (6,0) and (7,0) tubes (see Figs. 13, 14). It is also observed that the kinetic energy at the centre of the second tube drops relatively fast in the (6,0) tube in comparison to the (7,0) tube.

Heat pulse propagation is completely different for (8,0) and higher chirality tubes. For the (8,0) tube, the heat pulse decays while propagating along the first tube and also broadens with time, a behavior which is completely different from that observed in low chirality tubes. At time t = 1.45 ps, the peak kinetic energy of the pulse is 0.68 eV, and decays to 0.42 eV at t = 3.95 ps, Fig. 15a. The pulse speed in the (8,0) tube is 19 km/s, which is lower than the pulse speed observed in lower chirality tubes (see Table 2). Oman and Srivastava [60] observed a similar decay in the pulse speed in zig-zag tubes for twisting phonon modes (TW) with speed ranging in 16–18 km/s. The pulse excited in the second tube also shows different characteristics, and exhibits a pulse shape which is more flat at the center, Fig. 15b. All these behaviors imply a dissipative nature to the pulse propagation in (8,0) tubes.

Heat pulse propagation in a (10,0) tube is much different from that previously discussed for tubes of low chirality. The high kinetic energy in the heat pulse in the first tube cannot be sustained and the pulse decays very quickly. At t = 2.45 ps, a clearly identifiable wave shape evolves from the dissipating heat pulse; the speed of this wave is 12.5 km/s, which corresponds to the second sound wave speed observed in zig-zag tubes in Ref. [60]. In transient experiments, at low temperatures and in structures with high purity, normal phonon scattering processes can

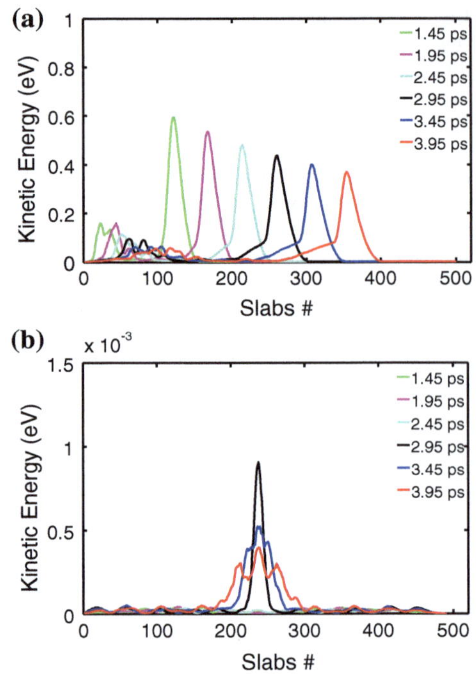

Fig. 15 **a** Location and shape of the heat pulse at different time instants along the first CNT for (8,0) chirality. The pulse generation starts at t = 0.0 ps. **b** Location and shape of the heat pulse at different time instants in the second CNT, also of the same chirality

play an important role. They may couple various phonon modes and make possible the collective oscillation in phonon density which is second sound. Second sound is observed under very restrictive conditions [60]. One of these conditions is that the momentum conserving normal phonon scattering processes should be dominant compared to the momentum randomizing Umklapp phonon scattering processes [60]. For CNTs, thermal conductivity increases even up to room temperature and it can reasonably be believed that N-process phonon interactions dominate over a wide range of temperature. These arguments and the observed speed of the wave (12.5 km/s = $1/3^{0.5}$ times the LA phonon speed) suggest that the observed wave pulse in the (10,0) nanotube corresponds to that of second sound.

The peak kinetic energy of the second sound wave mode observed in (10,0) tube is 0.12 eV, which is much smaller than the peak kinetic energy of leading heat pulses observed in the low-chiral tube configurations. In addition, due to the low speed, this pulse in the first CNT does not cross the contact zone in 4 ps (this is the time range for which we analyzed low-chirality tubes), so we have extended the analysis up to 6.5 ps. The energy gained by the second CNT due to the interaction with the heat pulse of the first CNT is very low. This is the reason for a very small rise in the peak kinetic energy ($\sim$0.2 meV) of the excited pulse in the second tube, Fig. 16b. The kinetic energy of the carbon atoms located at the contact area of the second CNT (black and red curves in Fig. 16b) is of the same order as that of the kinetic energy of atoms at other locations (see peaks at the CNT

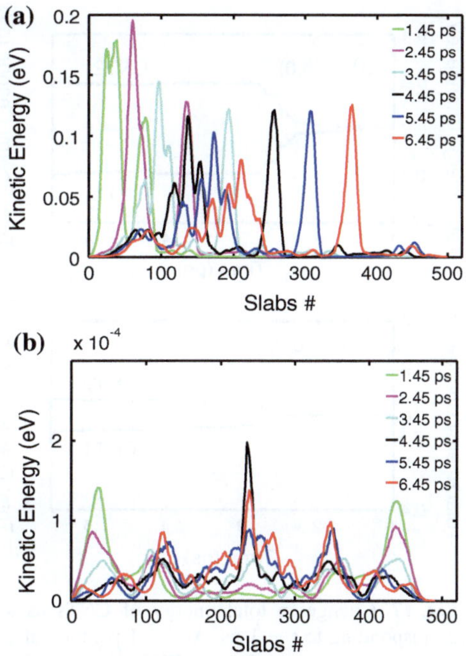

Fig. 16 **a** Location and shape of the heat pulse at different time instants along the first CNT for (10,0) chirality. The pulse generation starts at t = 0.0 ps. **b** Location and shape of the heat pulse at different time instants in the second CNT, also of the same chirality

ends in Fig. 16b). From our analysis, we have observed that in low-chirality CNTs we can generate a purely ballistic pulse, while in high-chirality CNTs, a diffusive tail is also present and the pulse cannot sustain its peak temperature during propagation. We see a difference between our results and Osman and Srivastava's results only for low-chirality CNTs.

The change in total energy of the CNTs as a function of time is plotted in Fig. 17 for (6,0), (7,0), (8,0) and (10,0) tubes. The change in total energy is computed with respect to the reference total energy at t = 1 ps; this is time at which the pulse generated in the first CNT starts propagating towards the contact region of the two CNTs. The magnitude of total energy exchange or increase in total energy of the second tube up to time instant 5.5 ps is presented in Table 2 for tubes of different chiralities. The energy exchange between the tubes is largest for the (6,0) tube and decreases for high-chirality tubes, Fig. 17. However a definitive statement regarding this cannot be made based on the present analysis as the energy exchange for the (5,0) tube is lower than that for the (6,0) tube by 0.07 eV and the energy exchange for the (7,0) tube is lower than that for the (8,0) tube by 0.028 eV. The above analysis shows that lower-chirality tubes have better coupling in comparison to high-chirality tubes as far as these heat pulses are concerned.

Phonon modes with high speed are very inefficient in transferring energy to the second tube as they spend very little time in the contact zone. Thus, it is likely that for realistic applications, slow-moving phonon modes (for example, optical modes) in the first CNT would be better-coupled to the second CNT. Shiomi and

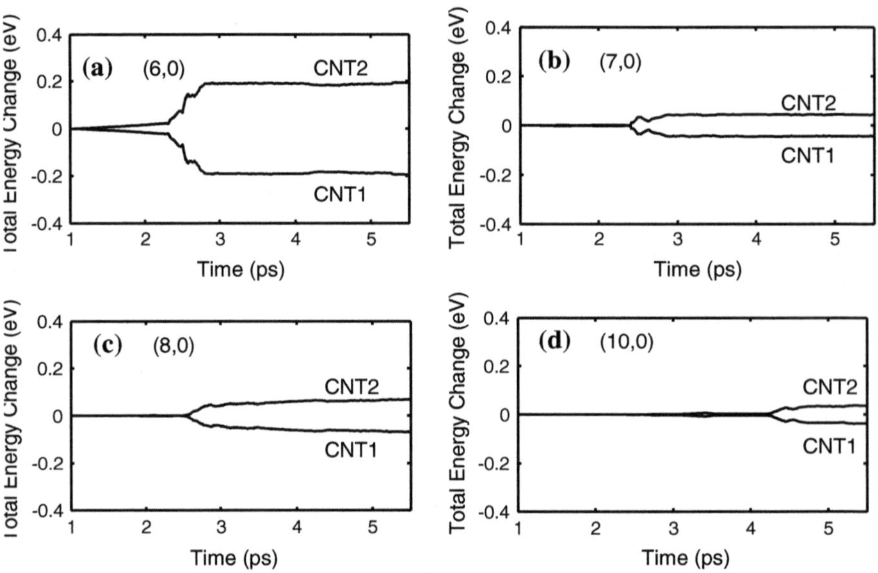

Fig. 17 Change in total energy of CNTs as a function of time with reference total energy corresponding to t = 1 ps. At t = 1 ps, the pulse generated in CNT-1 starts propagating towards the contact region of the two CNTs. Here, CNT-1 corresponds to the CNT in which the pulse is generated; CNT-2 is located in a crossed-configuration with respect to CNT-1. The chirality of the CNTs is **a** (6,0); **b** (7,0); **c** (8,0) and **d** (10,0)

Table 2 Pulse speed in the first tube and the maximum total energy rise in the second tube for different chiralities

CNT-chirality	Pulse speed (km/s)	Increase in total energy of second tube until t = 5.5 ps (eV)
(5,0)	22.0	0.129
(6,0)	23.2	0.203
(7,0)	22.6	0.046
(8,0)	19.0	0.074
(10,0)	12.5	0.037

Maruyama [62] observe from their modal analyses on a single tube using wavelet transformations that the major contribution to non-Fourier heat conduction comes from optical phonon modes with sufficient group velocity and with wave vectors in the intermediate regime for short nanotubes. Dispersion curves for CNTs of chirality (5,5) show that even at low frequencies, longitudinal and transverse optical modes may be present, with velocities comparable with the acoustic modes. In a nanotube network, the link between two nanotubes is of the order of just few nanometers. Therefore, in nanotube networks where the contact between the tubes governs the transport, optical phonon modes may be a dominant heat transfer pathway for communication between tubes.

5.2 Wavelet Analysis of Heat Pulse

Wavelet analysis of the heat pulse excited in the CNT configuration described in the previous section (see Fig. 12) is performed for different chiralities of the tubes. The wavelet transform (WT) is an analysis tool well-suited for the study of processes which occur over finite spatial and temporal domains. The wavelet transform is a generalized form of the Fourier transform (FT). A WT uses generalized local functions known as wavelets which can be stretched and translated with a desired resolution in both the frequency and time domains [63, 64].

Wavelets decompose a time series in the time-frequency space and are useful for identifying the evolution of dominant frequency modes with time. A time signal *s(t)* is decomposed using wavelet methods in terms of the elementary function $\psi_{b,a}$ derived from a mother wavelet ψ by dilation and translation [63]:

$$\psi_{b,a}(t) = \frac{1}{a^{0.5}}\psi\left(\frac{t-b}{a}\right). \tag{10}$$

Here, a and b are parameters which control dilation and translation respectively. The a parameter is also known as the *scale* in wavelet analysis. $\psi_{b,a}$ is known as the *daughter* wavelet as it is derived from the mother wavelet ψ using translation and dilation. The normalization factor $a^{0.5}$ ensures that the mother and daughter wavelets have the same energy. The wavelet transform (WT) of a signal *s(t)* is given as the convolution integral of *s(t)* with ψ^*, where ψ^* is the complex conjugate of the wavelet function ψ:

$$W(b,a) = \frac{1}{a^{0.5}}\int \psi^*\left(\frac{t-b}{a}\right)s(t)dt. \tag{11}$$

In general, wavelet functions are complex functions, so the WT are also complex, and have a real part, an imaginary part and a phase angle. The power spectrum of a WT is defined as $|W|^2$. We use the Morlet wavelet for heat pulse analysis; this wavelet has the form of a plane wave with a Gaussian envelope [63]. The Morlet wavelet is given by:

$$\psi = \mathrm{e}^{iwt} \times \mathrm{e}^{-0.5|t|^2}. \tag{12}$$

The form of this wavelet is shown in Fig. 18 after translating by different values of b and dilating by different values of the scale parameter a.

The power spectrum of the velocity magnitude of each atom in the nanotube is computed using the method described above. By summing the power spectrum over all the atoms of one ring, a one-dimensional projection of the temporal spectra along the nanotube axis is obtained. In this way, temporally evolving spectra of the velocity magnitude for the entire spatio-temporal field are obtained.

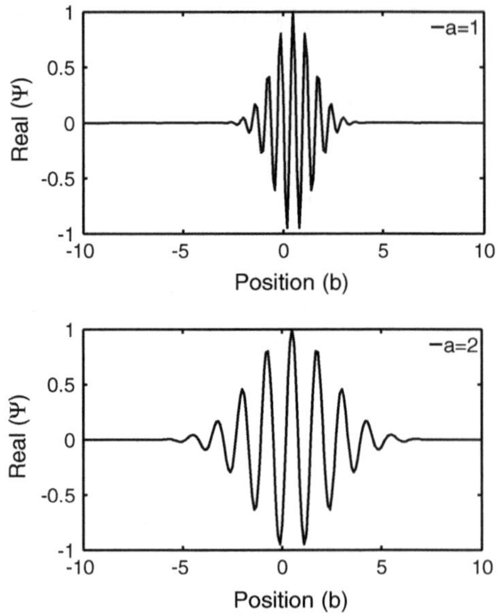

Fig. 18 An example of a Morlet wavelet with different values of the scale *a* [63]

Wavelet analysis of the energy modes excited in the second CNT due to the interaction with first CNT is performed for chirality of tubes varying from (5,0) to (10,0). The power spectrum of the velocity magnitude of each atom in the nanotube is computed using the method described above. Temporally evolving spectra of the velocity magnitude for the entire spatio-temporal field are computed. Results are presented as temporal sequences of the spectral contours in the frequency-space domain in Fig. 19 for a (5,0) CNT. Six temporal sequences are used, which correspond to $t = 2.0$, 2.5, 3.0, 3.5, 4.0 and 5.0 ps. In these plots, the vertical axis represents the frequency in THz and the horizontal axis represents the spatial location along the nanotube axis in terms of the slab (i.e. ring) numbers.

The evolution and propagation of spectral modes along the tube axis for a (5,0) CNT at different time instants may be observed in Fig. 19. At $t = 2$ ps, the heat pulse in the first tube is approaching the contact zone, Fig. 19a; thus some low frequency phonon modes are excited in the second nanotube at the location of the contact. At $t = 2.5$ ps (Fig. 19b), the heat pulse in the first tube crosses the contact zone; this is the time at which the peak temperature of the thermal pulse excited in the second CNT is observed (Fig. 13b). Phonon modes of frequencies up to 30–35 THz are excited in the second nanotube by this time, but most of the dominant frequencies are less than 10 THz, Fig. 19b. Once the heat pulse in the first tube moves away from the contact zone, the energy given to the second tube spreads to both sides of the tube symmetrically. Different phonon modes are excited along the second tube, and can be observed from the contour plots in Fig. 19c–f. The contour plot at $t = 3.0$ ps shows two distinctly visible modes, one centered around 5 THz and another centered around 10 THz. The speed of these modes may be computed from the movement of the high power spectrum color patches

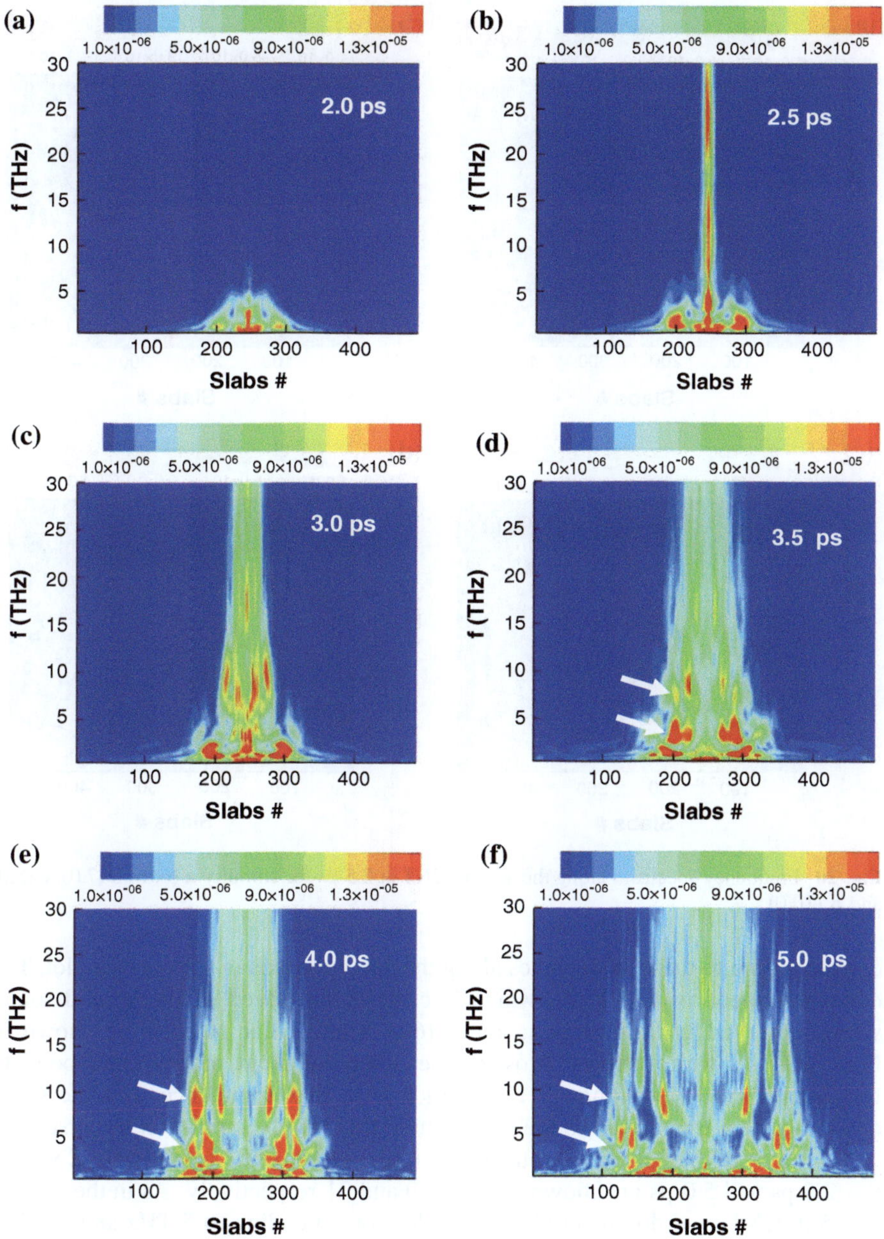

Fig. 19 Frequency spectrum along the second nanotube for (5,0) chirality at different time instants. **a** 2.0 ps, **b** 2.5 ps, **c** 3.0 ps, **d** 3.5 ps, **e** 4.0 ps, **f** 5.0 ps. *White arrows* shows frequencies corresponding to 5 and 10 THz

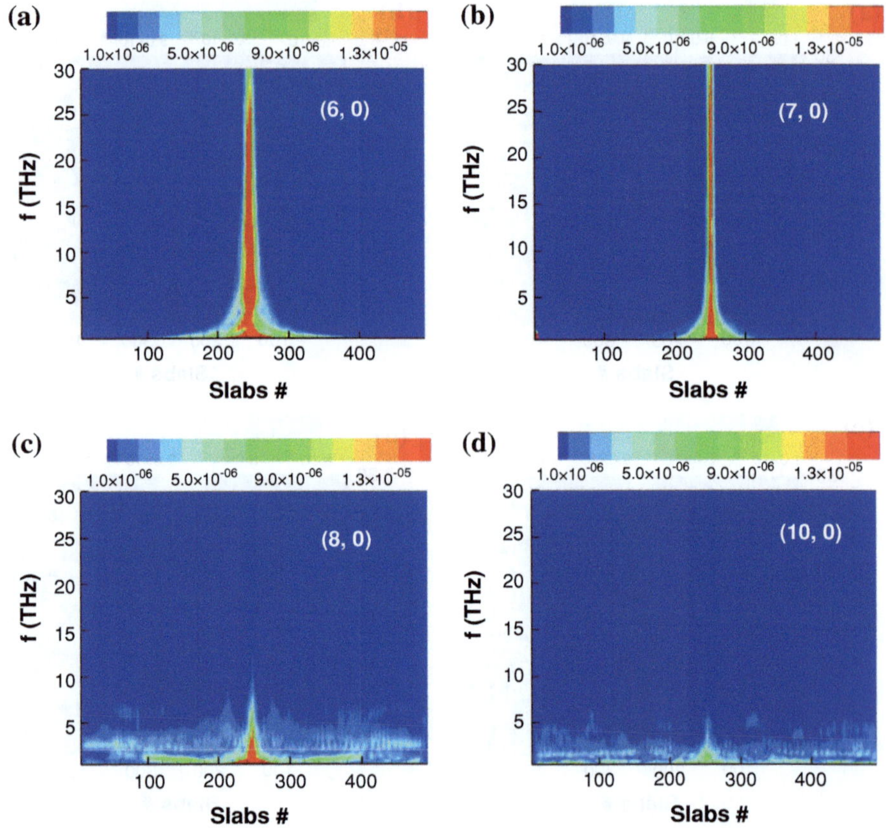

Fig. 20 Frequency spectrum along the second CNT at 2.5 ps for chirality **a** (6,0); **b** (7,0); **c** (8,0) and **d** (10,0)

corresponding to these frequencies along the nanotube axis in the contour plot. The computed speed is 12.0 km/s, which is close to the speed of the second sound waves observed in these zig-zag tubes [60]. These color patches are shown in Fig. 19d–f by white arrows. Most of the other dominant modes correspond to frequencies even lower than 5 THz, which is in correspondence with the small amount of heat transferred to the second tube from the heat pulse in the first tube.

Contours of the power spectrum for (6,0), (7,0), (8,0) and (10,0) CNTs at $t = 2.5$ ps and 5.0 ps are shown in Figs. 20 and 21 respectively. As in the case of the (5,0) CNT, two dominant phonon modes corresponding to 5 THz and 21 THz are seen to propagate along the second tube for (6,0) and (7,0) CNTs, Fig. 21a, b. These contour plots also clearly show that the rate of heat spreading along the CNT axis slows down with increasing diameter. This is also observed in the heat pulse analysis in the previous section by examining the location and shape of the heat pulses generated in the second tube. A similar comparison cannot be made for (8,0) and (10,0) CNTs because the pulse generated in the first CNT is not sustained

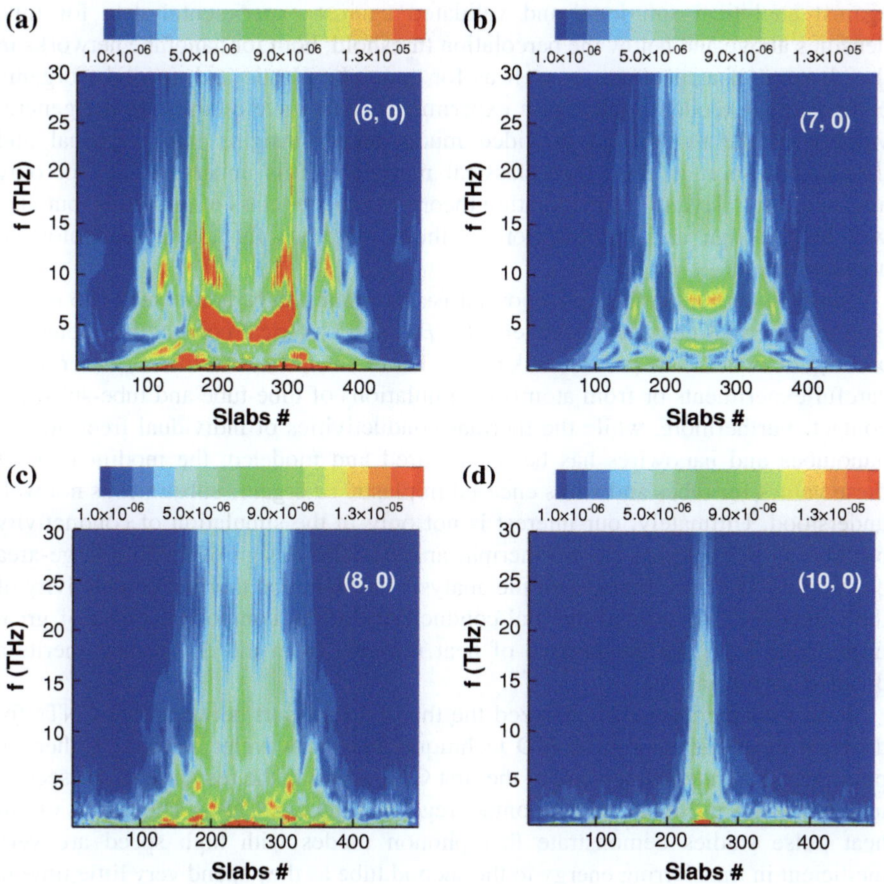

Fig. 21 Frequency spectrum along the second CNT at 5.0 ps for chirality **a** (6,0); **b** (7,0); **c** (8,0) and **d** (10,0)

at its initial peak temperature, and very little heat is transferred to the second CNT for these cases. The spectral modes for the (10,0) CNT are completely different from other CNTs; most of the high energy modes present in low-chirality CNTs are not excited, Fig. 21d. This is due to the evolution of second sound waves in the first CNT; the peak temperature is an order of magnitude lower than that in low chirality CNTs, Figs. 16b, 21d.

6 Conclusions

In summary, a computational model for thermal and electrical transport in nanotube composites has been developed and applied to the determination of lateral electrical and thermal conductivity of finite thin films. The model has been verified against analytical solutions, and validated against experimental data for tube densities above and below the percolation threshold, both for nanotube networks in the absence of a substrate, as well as for nanotube-plastic and nanotube-organic composites. Excellent match with experimental data have established the general validity of the model and provided much deeper insights into electrical and thermal transport in these non-classical materials. This understanding not only advances the relevance of percolation theory to broader class of problems, but also will help design and optimization of these materials for future technological applications.

Nevertheless, a number of important issues remain to be addressed. The model employs thermal contact parameters Bi_c, Bi_s, and electrical contact parameters c_{ij} and d_{is}, which are at present unknown. These must be determined either from careful experiments or from atomistic simulations of tube-tube and tube-substrate contact. Furthermore, while the thermal conductivities of individual freestanding nanotubes and nanowires has been measured and modeled, the modification of these values for tubes and wires encased in plastic or organic substrates is not well understood. Ultimately, our interest is not only in the simulation of conductivity but also in the coupled electro-thermal analysis and design of flexible large-area electronics. We have focused on the analysis of the lateral thermal conductivity of the composite, but normal thermal conductivity of the composite is also of great importance from the perspective of heat removal from the device and merits a detailed investigation.

In this chapter, we also analyzed the thermal transport between two CNTs for different chiralities using the MD technique and the wavelet method. A thermal pulse was generated at one end of the first CNT to study the response of the second tube when this pulse passed the contact region of the two CNTs. The results of our heat pulse studies demonstrate that phonon modes with high speed are very inefficient in transferring energy to the second tube as they spend very little time in the contact zone. Most likely, slow-moving phonon modes in the first CNT would be better coupled to the second CNT and would dominate thermal boundary conductance between these CNTs. To better understand the coupling modes between tubes, a wave packet corresponding to specific phone modes should be passed through a CNT and its interaction with other CNTs should be studied. Understanding the coupling between CNTs of different chirality and contact area could also be of great interest for engineering the properties of CNT-network based composites. Most of the present work is in the low temperature regime. It is also important to investigate transport mechanisms in the high temperature regime using either appropriate numerical modeling or experimental techniques.

Acknowledgments Support of J. Murthy and S. Kumar under NSF grants CTS-0312420, CTS-0219098, EE-0228390, the Purdue Research Foundation and Purdue's Network for Computational Nanotechnology (NCN) is gratefully acknowledged.

References

1. Hur, S.H., Kocabas, C., Gaur, A., et al.: Printed thin-film transistors and complementary logic gates that use polymer-coated single-walled carbon nanotube networks. J. Appl. Phys. **98**(11), 114302 (2005)
2. Reuss, R.H., Chalamala, B.R., Moussessian, A., et al.: Macroelectronics: perspectives on technology and applications. Proc. IEEE **93**(7), 1239–1256 (2005)
3. Collins, P.C., Arnold, M.S., Avouris, P.: Engineering carbon nanotubes and nanotube circuits using electrical breakdown. Science **292**(5517), 706–709 (2001)
4. Kagan, C.R., Andry, P.: Thin film transistors. Marcel Dekker, New York (2003)
5. Novak, J.P., Snow, E.S., Houser, E.J., et al.: Nerve agent detection using networks of single-walled carbon nanotubes. Appl. Phys. Lett. **83**(19), 4026–4028 (2003)
6. Alam, M.A., Nair, P.R.: Geometry of diffusion and the performance limits of nanobiosensors. Nanotechnology 501 lecture series. https://www.nanohub.org/resources/2048/(2006)
7. Madelung, O.: Technology and applications of amorphous silicon. Springer, Berlin (2000)
8. Zhou, Y.X., Gaur, A., Hur, S.H., et al.: P-channel, n-channel thin film transistors and p-n diodes based on single wall carbon nanotube networks. Nano Lett. **4**(10), 2031–2035 (2004)
9. Snow, E.S., Campbell, P.M., Ancona, M.G., et al.: High-mobility carbon-nanotube thin-film transistors on a polymeric substrate. Appl. Phys. Lett. **86**(3), 066802 (2005)
10. Snow, E.S., Novak, J.P., Lay, M.D., et al.: Carbon nanotube networks: nanomaterial for macroelectronic applications. J. Vac. Sci. Technol. B **22**(4), 1990–1994 (2004)
11. Snow, E.S., Novak, J.P., Campbell, P.M., et al.: Random networks of carbon nanotubes as an electronic material. Appl. Phys. Lett. **82**(13), 2145–2147 (2003)
12. Cao, Q., Rogers, J.A.: Ultrathin films of single-walled carbon nanotubes for electronics and sensors: a review of fundamental and applied aspects. Adv. Mater. **21**(1), 29–53 (2009)
13. Kumar, S., Murthy, J.Y., Alam, M.A.: Percolating conduction in finite nanotube networks. Phys. Rev. Lett. **95**(6), 066802 (2005)
14. Hur, S.H., Khang, D.Y., Kocabas, C., et al.: Nanotransfer printing by use of noncovalent surface forces: applications to thin-film transistors that use single-walled carbon nanotube networks and semiconducting polymers. Appl. Phys. Lett. **85**(23), 5730–5732 (2004)
15. Kocabas, C., Hur, S.H., Gaur, A., et al.: Guided growth of large-scale, horizontally aligned arrays of single-walled carbon nanotubes and their use in thin-film transistors. Small **1**(11), 1110–1116 (2005)
16. Kocabas, C., Shim, M., Rogers, J.A.: Spatially selective guided growth of high-coverage arrays and random networks of single-walled carbon nanotubes and their integration into electronic devices. J. Am. Chem. Soc. **128**(14), 4540–4541 (2006)
17. Milton, G.W.: The Theory of Composites. Cambridge University Press, New York (2002)
18. Nan, C.W., Birringer, R., Clarke, D.R., et al.: Effective thermal conductivity of particulate composites with interfacial thermal resistance. J. Appl. Phys. **81**(10), 6692–6699 (1997)
19. Kumar, S., Alam, M.A., Murthy, J.Y.: Computational model for transport in nanotube-based composites with applications to flexible electronics. ASME J. Heat Transf. **129**(4), 500–508 (2007)
20. Kumar, S., Pimparkar, N., Murthy, J.Y., et al.: Theory of transfer characteristics of nanotube network transistors. Appl. Phys. Lett. **88**, 123505 (2006)
21. Zhang, G., Qi, P., Wang, X., et al.: Selective etching of metallic carbon nanotubes by gas-phase reaction. Science **314**, 974–977 (2006)
22. Arnold, M.S., Green, A.A., Hulvat, J.F., et al.: Sorting carbon nanotubes by electronic structure using density differentiation. Nat. Nanotechnol. **1**, 60–65 (2006)
23. Huang, H., Liu, C., Wu, Y., et al.: Aligned carbon nanotube composite films for thermal management. Adv. Mater. **17**, 1652 (2005)
24. Nan, C.W., Liu, G., Lin, Y.H., et al.: Interface effect on thermal conductivity of carbon nanotube composites. Appl. Phys. Lett. **85**(16), 3549–3551 (2004)

25. Biercuk, M.J., Llaguno, M.C., Radosavljevic, M., et al.: Carbon nanotube composites for thermal management. Appl. Phys. Lett. **80**(15), 2767–2769 (2002)
26. Xu, X.J., Thwe, M.M., Shearwood, C., et al.: Mechanical properties and interfacial characteristics of carbon-nanotube-reinforced epoxy thin films. Appl. Phys. Lett. **81**(15), 2833–2835 (2002)
27. Reibold, M., Paufler, P., Levin, A.A., et al.: Carbon nanotubes in an ancient damascus sabre. Nature **444**(16), 286 (2006)
28. Hu, L., Hecht, D.S., Gruner, G.: Percolation in transparent and conducting carbon nanotube networks. Nano Lett. **4**(12), 2513–2517 (2004)
29. Keblinski, P., Cleri, F.: Contact resistance in percolating networks. Phys. Rev. B **69**(18), 184201 (2004)
30. Hu, T., Grosberg, A.Y., Shklovskii, B.I.: Conductivity of a suspension of nanowires in a weakly conducting medium. Phys. Rev. B **73**(15), 155434 (2006)
31. Lukes, J.R., Zhong, H.L.: Thermal conductivity of individual single-wall carbon nanotubes. J. Heat Transf. Trans. ASME **129**(6), 705–716 (2007)
32. Maruyama, S., Igarashi, Y., Shibuta, Y.: Molecular dynamics simulations of heat transfer issues in carbon nanotubes. In: The 1st International Symposium on Micro and Nano Technology. Honolulu, Hawaii, USA (2004)
33. Small, J.P., Shi, L., Kim, P.: Mesoscopic thermal and thermoelectric measurements of individual carbon nanotubes. Solid State Commun. **127**(2), 181–186 (2003)
34. Maune, H., Chiu, H.Y., Bockrath, M.: Thermal resistance of the nanoscale constrictions between carbon nanotubes and solid substrates. Appl. Phys. Lett. **89**(1), 013109 (2006)
35. Carlborg, C.F., Shiomi, J., Maruyama, S.: Thermal boundary resistance between single-walled carbon nanotubes and surrounding matrices. Phys. Rev. B **78**(20), 205406 (2008)
36. Zhong, H.L., Lukes, J.R.: Interfacial thermal resistance between carbon nanotubes: molecular dynamics simulations and analytical thermal modeling. Phys. Rev. B **74**(12), 125403 (2006)
37. Greaney, P.A., Grossman, J.C.: Nanomechanical energy transfer and resonance effects in single-walled carbon nanotubes. Phys. Rev. Lett. **98**(12), 125503 (2007)
38. Prasher, R.S., Hu, X.J., Chalopin, Y., et al.: Turning carbon nanotubes from exceptional heat conductors into insulators. Phys. Rev. Lett. **102**, 105901 (2009)
39. Pimparkar, N., Kumar, S., Murthy, J.Y., et al.: Current-voltage characteristics of long-channel nanobundle thin-film transistors: A 'bottom-up' perspective. IEEE Electron Device Lett. **28**(2), 157–160 (2006)
40. Bo, X.Z., Lee, C.Y., Strano, M.S., et al.: Carbon nanotubes-semiconductor networks for organic electronics: the pickup stick transistor. Appl. Phys. Lett. **86**(18), 182102 (2005)
41. Kumar, S., Blanchet, G.B., Hybertsen, M.S., et al.: Performance of carbon nanotube-dispersed thin-film transistors. Appl. Phys. Lett. **89**(14), 143501 (2006)
42. Kundert, K.S.: Sparse user's guide. Department of Electrical Engineering and Computer Sciences, University of California, Berkeley (1988)
43. Pike, G.E., Seager, C.H.: Percolation and conductivity—computer study 1. Phys. Rev. B **10**(4), 1421–1434 (1974)
44. Alam, M.A.: Nanostructured electronic devices: percolation and reliability. Intel-purdue summer school on electronics from bottom up. http://nanohub.org/resources/7168 (2009)
45. Foygel, M., Morris, R.D., Anez, D., et al.: Theoretical and computational studies of carbon nanotube composites and suspensions: electrical and thermal conductivity. Phys. Rev. B **71**(10), 104201 (2005)
46. Foygel, M., Morris, R.D., Anez, D., et al.: Theoretical and computational studies of carbon nanotube composites and suspensions: electrical and thermal conductivity. Phys. Rev. B **71**(10), 104201 (2005)
47. Shenogina, N., Shenogin, S., Xue, L., et al.: On the lack of thermal percolation in carbon nanotube composites. Appl. Phys. Lett. **87**(13), 133106 (2005)
48. Frank, D.J., Lobb, C.J.: Highly efficient algorithm for percolative transport studies in 2 dimensions. Phys. Rev. B **37**(1), 302–307 (1988)

49. Lobb, C.J., Frank, D.J.: Percolative conduction and the alexander-orbach conjecture in 2 dimensions. Phys. Rev. B **30**(7), 4090–4092 (1984)
50. Pimparkar, N., Alam, M.A.: A "Bottom-up" redefinition for mobility and the effect of poor tube-tube contact on the performance of CNT nanonet thin-film transistors. IEEE Electron Device Lett. **29**(9), 1037–1039 (2008)
51. Taur, Y., Ning, T.: Fundamentals of modern VLSI devices. Cambridge University Press, New York (1998)
52. Fuhrer, M.S., Nygard, J., Shih, L., et al.: Crossed nanotube junctions. Science **288**(5465), 494–497 (2000)
53. Seidel, R.V., Graham, A.P., Rajasekharan, B., et al.: Bias dependence and electrical breakdown of small diameter single-walled carbon nanotubes. J. Appl. Phys. **96**(11), 6694–6699 (2004)
54. Huxtable, S.T., Cahill, D.G., Shenogin, S., et al.: Interfacial heat flow in carbon nanotube suspensions. Nat. Mater. **2**(11), 731–734 (2003)
55. Kumar, S., Alam, M.A., Murthy, J.Y.: Effect of percolation on thermal transport in nanotube composites. Appl. Phys. Lett. **90**(10), 104105 (2007)
56. Bryning, M.B., Milkie, D.E., Islam, M.F., et al.: Thermal conductivity and interfacial resistance in single-wall carbon nanotube epoxy composites. Appl. Phys. Lett. **87**(16), 161909 (2005)
57. Hung, M.T., Choi, O., Ju, Y.S., et al.: Heat conduction in graphite-nanoplatelet-reinforced polymer nanocomposites. Appl. Phys. Lett. **89**(2), 023117 (2006)
58. Kumar, S., Murthy, J.Y.: Interfacial thermal transport between nanotubes. J. Appl. Phys. **106**(8), 084302 (2009)
59. Brenner, D.W., Shenderova, O.A., Harrison, J.A., et al.: A second-generation reactive empirical bond order (rebo) potential energy expression for hydrocarbons. J. Phys. Condens. Matter **14**(4), 783–802 (2002)
60. Osman, M.A., Srivastava, D.: Molecular dynamics simulation of heat pulse propagation in single-wall carbon nanotubes. Phys. Rev. B **72**(12), 125413 (2005)
61. Erhart, P., Albe, K.: The role of thermostats in modeling vapor phase condensation of silicon nanoparticles. Appl. Surf. Sci. **226**(1–3), 12–18 (2004)
62. Shiomi, J., Maruyama, S.: Non-fourier heat conduction in a single-walled carbon nanotube: Classical molecular dynamics simulations. Phys. Rev. B **73**(20), 205420 (2006)
63. Lau, K.M., Weng, H.: Climate signal detection using wavelet transform: How to make a time series sing. Bull. Am. Meteorol. Soc. **76**(12), 2391–2402 (1995)
64. Torrence, C., Compo, G.P.: A practical guide to wavelet analysis. Bull. Am. Meteorol. Soc. **79**(1), 61–78 (1998)
65. Liu, C.H., Huang, H., Wu, Y., et al.: Thermal conductivity improvement of silicone elastomer with carbon nanotube loading. Appl. Phys. Lett. **84**(21), 4248–4250 (2004)
66. Gong, Q.M., Li, Z., Bai, X.D., et al.: Thermal properties of aligned carbon nanotube/carbon nanocomposites. Mater. Sci. Eng. a-Struct. Mater. Prop. Microstruct. Process. **384**(1–2), 209–214 (2004)
67. Choi, S.U.S., Zhang, Z.G., Yu, W., et al.: Anomalous thermal conductivity enhancement in nanotube suspensions. Appl. Phys. Lett. **79**(14), 2252–2254 (2001)
68. Wen, D.S., Ding, Y.L.: Effective thermal conductivity of aqueous suspensions of carbon nanotubes (carbon nanotubes nanofluids). J. Thermophys. Heat Transf. **18**(4), 481–485 (2004)

Thermo-Mechanical Modelling of High Energy Particle Beam Impacts

M. Scapin, L. Peroni and A. Dallocchio

Abstract The unprecedented energy intensities of modern hadron accelerators yield special problems with the materials that are placed close to or into the high intensity beams. The energy stored in LHC in a single beam is equivalent to about 80 kg of TNT explosive, stored in a transverse beam area of 0.2 mm × 0.2 mm. The materials placed close to the beam are used at, or even beyond, their damage limits. However, it is very difficult to predict structural efficiency and robustness accurately: beam-induced damage occurs in a regime where practical experience does not exist. This study is performed in order to estimate the damage on a copper component due to the impact with a 7 TeV proton beam generated by LHC. The case study represents an accidental case consequent to an abnormal release of the beam, in which 8 bunches irradiate the target directly. The energy delivered on the component is calculated using the FLUKA code and then used as input in the numerical simulations, that are carried out via the FEM code LS-DYNA. Different numerical models are realized trying to obtain the simplest model able to correctly describe the material response without affecting the goodness of the results.

Keywords CERN · Equation of state · High energy impact · LS-DYNA · Shockwaves

M. Scapin (✉) · L. Peroni
Politecnico di Torino, Corso Duca degli Abruzzi, 24 10129 Torino, Italy
e-mail: martina.scapin@polito.it

L. Peroni
e-mail: lorenzo.peroni@polito.it

A. Dallocchio
Mechanical and Materials Group, Engineering, CERN, 1211 Geneva 23, Switzerland
e-mail: alessandro.dallocchio@cern.ch

N. Bonora and E. Brown (eds.), *Numerical Modeling of Materials Under Extreme Conditions*, Advanced Structured Materials 35, DOI: 10.1007/978-3-642-54258-9_4,

1 Introduction

Particle accelerators [1] act as microscopes for such a complex research; these large machines accelerate charged elementary particles (electrons, protons or ionized atoms) to high kinetic energies. A high energy particle beam can be brought into collision against a fixed target or against another beam and from this encounter a multitude of short life sub-atomic particles is originated. The investigation of matter in these extreme conditions can be compared with the status of the universe in the first moments after the so-called 'Big Bang'. The higher the energy of the colliding beams and the event rate, the wider the spectrum of the generable sub-atomic particles. It is in this perspective that the project of building the Large Hadron Collider (LHC) [2] at the European Organization for Nuclear Research (CERN, Geneva) was approved.

The LHC [3] is a circular accelerator with a 26.659 km circumference situated at the border between Switzerland and France at an average depth of 100 m underground. This machine mainly provides the collision between two counter-circulating proton beams. Each proton beam consists of 3×10^{14} protons at 7 TeV, so when the protons collide the collision energy is 14 TeV. Under nominal operating conditions, the beam has 2,808 bunches each having 1.15×10^{11} protons. The bunch length is 0.5 ns and the time between two successive bunches is 25 ns, so the duration of the entire beam is about 72 μs.

The total energy stored in each beam at maximum energy is about 350 MJ, two orders of magnitude beyond the achievements in the Tevatron or HERA [4]. This is enough energy to melt 500 kg of copper.

This large amount of energy is potentially destructive for accelerator equipments having direct interaction with particles in case of uncontrolled beam loss, so everything is done to ensure that this never happens. Besides, it is important to know what will be the damage in case of the LHC malfunction. It is in this perspective that a thermo-mechanical analysis becomes relevant. However, it is very difficult to predict structural efficiency and robustness accurately: beam-induced damage for high energy and high intensity occurs in a regime where practical experience does not exist.

When a particle beam interacts with a solid target the particles deposit their energy on the hit material. So they provoke a dynamic response of the structure entailing thermal stress waves and thermally induced vibrations or even the failure of the component. The pressure and temperature increase and the materials could arrive at its melting temperature or vaporize. The remaining part of the material is characterized by high values of plastic strain, strain-rate and temperature.

From these considerations it is clear what is the complexity of the problem: in order to correctly simulate the thermo-mechanical response of the hit material it is needed to take into account both the hydrodynamic behaviour using a dedicated equation of state (EOS) and the deviatoric behaviour using a dedicated material model.

The numerical simulations are performed using the commercial FEM code LS-DYNA [5]. For the simulations the chosen equation of state is a polynomial EOS, in which the coefficients are obtained fitting a three-phase tabular equation of state, and the material model is the Johnson–Cook model.

The evaluation of thermal loads on the hit material is performed by the Fluka Team at CERN using a statistical code, called FLUKA [6], based on the Monte-Carlo method. FLUKA returns an energy map on a particular geometry taking into account all the particles in the cascade generated by the interaction between the proton beam and the target. Finally the FLUKA results are used as input for thermo-structural studies.

As mentioned before, the material involves in such high energy and high intensity impacts operates under extreme conditions, in which the possibility to perform experimental tests is limited. For this reason the importance of developing reliable methods and accurate models that could be efficiently applied to estimate the damage occurring during an impact is therefore evident.

2 State of Art

Before describing the implemented methods, of course, a short discussion about the approaches developed by other research centres is presented. As a matter of fact, in literature different approaches to solve similar problems can be found.

In [7] the study of the thermo-mechanical effects induced in solids by high energy particles beams is carried out via two different approaches: an analytical method and a numerical one. Regarding of the numerical approach, the thermo-structural analysis is performed using the commercial code ANSYS. The starting point of the analysis is the energy map on the component calculated using FLUKA code, which is converted in a thermal power distribution used as input for the solution of the thermal problem, so the temperature distribution is calculated as function of space and time. The results of temperature analysis are used as nodal loads for the structural analysis, so dynamic thermal stresses and displacements are evaluated. Finally, a quasi-static step is performed in order to calculate potential permanent deformations of the structure.

A similar approach is carried out in [8, 9], in which the energy calculation on the material resulting from high energy particles beam impact with a solid material is performed via MARS code [10]. From the energy map the temperature rise is calculated and used in LS-DYNA (for an explicit analysis) or ANSYS (for an implicit analysis) code to calculate the dynamic stresses in the target.

In these two approaches the calculation of the temperature from the energy map is performed without using an equation of state, but simply considering the transformation in a solid material and using as conversion factor the heat capacity at constant volume for the solid.

It is important to underline that since the energy levels involved are very high, the material could undergo changes of state, so the use of an equation of state is needed in order to correctly evaluate the state of the material.

In this sense there are [11, 12], in which the energy deposition is evaluated via FLUKA code and the energy is directly used as input in the hydrodynamic code BIG-2 in order to calculate pressure, temperature and density of the hit material.

To do this a sophisticate three-phase equation of state is used. On the other hand, the approach developed in [11, 12] does not take into account the deviatoric part of the stress that is indeed negligible in the part of the material closest to the beam impact point, while it is by no means negligible in the remaining portion of material.

The purpose of this study is to analyze simultaneously the two aspects: the deviatoric behaviour and the hydrodynamic one, using the FEM code LS-DYNA in which there is the possibility to insert an equation of state. In this sense a previous study is developed in [13], in which a coupled mechanical, thermal and electromagnetic simulation is performed. So, the problem is quite different, but the approach can be similar. In fact, also in [13] the problem is linked with a rise in temperature that can melt or vaporize a portion of material. The difference is the cause of the heating: the temperature does not rise due to the impact of a high energy particle beam, but due to the Joule effect induced in material by a current. The simulations are performed via LS-DYNA code in which there is the possibility to implement an user-defined EOS. In [13] the three-phase equation of state GRAY is implemented and the input energy distribution comes from the electromagnetic solver.

3 Interaction Between a Particle Beam and a Solid Target

As mentioned in the introduction, the energy stored in LHC is unprecedented and potentially destructive for the accelerator equipments having direct interaction with high energy particles. High energetic beams may interact with several components present in particle accelerators, so it is of primary importance to predict the extension of the possible damage.

Different thermal loads must be taken into account depending on the interaction condition. In normal situations, a continuous interaction provokes a constant energy-rate deposition over a long period (from some second up to few hours). In this case thermal stresses and deformations take place but no dynamic response can usually be observed. Otherwise, in case of abnormal beam impacts, energy is rapidly deposited in time-scales of the order of microseconds or nanoseconds. This load condition typically entails a dynamic response of the structure. The resulting thermal stresses and deformations may affect the integrity or the proper functionality of the hit equipments. From this it is clear that an in-depth thermo-mechanical analysis is strongly necessary.

Interaction between high energy particle beams and solids can be considered from a structural point of view as an energy deposition inducing a sudden non uniform temperature increase. In function of which part of material is investigated the behaviour is different (see Fig. 1).

In the material part closest to the beam, the pressure and temperature increase and the materials could arrive at its melting temperature or vaporize. The material response in this condition is correctly described only using an equation of state that is able to describe the hydrodynamic behaviour, while in this portion of material

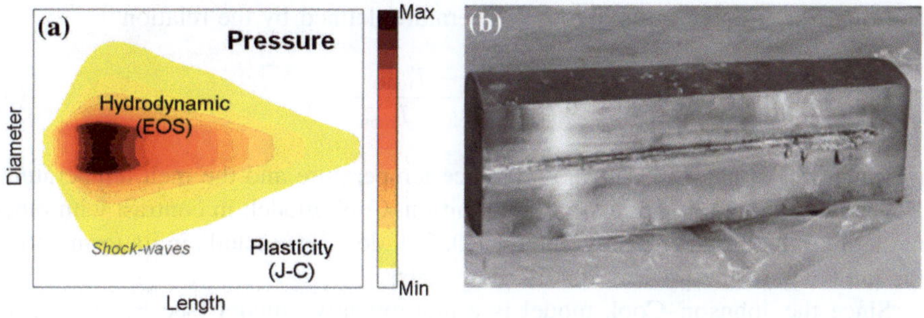

Fig. 1 Pressure distribution in case of impact between the high energy beam and a solid cylindrical target (**a**); damage on a metal target due to the impact (**b**)

the deviatoric stress is totally negligible. On the other hand, the remaining part of the material is characterized by high values of plastic strain, strain-rate and temperature, so the response is related with the strength material model used.

4 Equation of State and Material Model

As mentioned before, in order to completely describe the response of a material hit by a high energy particle beam both a strength material model and an equation of state are needed.

In the literature there are many material models that keep into account strain-rate and temperature influences on mechanical behaviour. Elasto-plastic models such those proposed by Johnson–Cook [14] and Cowper–Symonds [15] are the most used. In general in FEM codes there are also more complex models, such as Zerilli–Armstrong [16], but they request the identification of a larger set of parameters.

The material model chosen in these analyses is Johnson-Cook material model implemented in LS-DYNA code [5] and widely discussed in [17–19]. It is a purely empirical model that allows to take into account the effect of plastic strain, strain-rate and temperature. This model expresses the flow stress as

$$\sigma_y = \left(A + B\varepsilon_p^n\right)(1 + C\ln\dot{\varepsilon}^*)\left(1 - T^{*^m}\right) \tag{1}$$

where A, B and n are the material strain hardening constants, C is the strain-rate parameter, m is the temperature sensitivity, ε_p is the effective plastic strain, $\dot{\varepsilon}^*$ is the effective total strain rate normalized by quasi-static threshold rate ($\dot{\varepsilon}_0$) and defined as

$$\dot{\varepsilon}^* = \frac{\dot{\varepsilon}}{\dot{\varepsilon}_0} \tag{2}$$

and finally T^{*m} is the homologous temperature defined by the relation

$$T^* = \frac{T - T_{\text{room}}}{T_{\text{melt}} - T_{\text{room}}} \quad (3)$$

in which T_{room} and T_{melt} are the reference temperature and the melt temperature, respectively. It can be notice that the Johnson–Cook model, in contrast with other model like the Steinberg-Lund model [20, 21], does not include in its formulation the influence of the pressure on the flow stress.

Since the Johnson–Cook model is a multiplicative model (see Eq. 1) it considers the effects of strain, strain-rate and temperature independent, even if in the experimental dynamic tests the conversion of mechanical work into heat couples them and make them difficult separable. It is also important to note that when the temperature reaches the value of the melting temperature the mechanical strength of the material becomes zero. This means that the material loses its shear resistance and starts to be considered like a fluid.

The model is widely used in LS-DYNA calculations because it takes into account also the effect of the thermal softening, due to heat conversion of plastic work occurring at high strain-rate deformations. In fact, in this case the thermal diffusion can be neglected and the temperature rise can be evaluated under adiabatic assumption. Under this hypothesis and the further assumption of uniform stress the temperature can be analytically computed as a function of plastic work as

$$T = T_{\text{room}} + \frac{\beta}{\rho \cdot C_p} \cdot \int_0^{\varepsilon_p} \sigma(\varepsilon, \dot{\varepsilon}, T) \cdot d\varepsilon \quad (4)$$

where ρ is the material density, C_p is the specific heat at constant pressure and β is the Taylor-Quinney coefficient [22, 23] that represents the portion of plastic work converted into heat (for metals about 0.9).

In LS-DYNA the Johnson-Cook material model includes also a fracture model in which the strain at fracture is given by

$$\varepsilon_f = \left(D_1 + D_2 e^{D_3 \sigma^*}\right)(1 + D_4 \ln \dot{\varepsilon}^*)(1 + D_5 T^*) \quad (5)$$

where D_1, D_2, D_3, D_4 and D_5 are the failure parameters and σ^* is the ratio between the pressure and the effective stress. The material fracture occurs when the damage parameter

$$D = \sum \frac{\Delta \varepsilon_p}{\varepsilon_f} \quad (6)$$

reaches the value of 1. Besides the J–C material implemented in LS-DYNA offers also a spall model to represent the failure under hydrostatic tensile loads.

The other fundamental aspect to consider is the implementation of an equation of state to simulate the hydrodynamic material behaviour. An EOS is a constitutive relation between state variables and describes the state of the matter. Usually it expresses a thermodynamic variable (such as pressure) in function of two other independent state variables (such as density and internal energy).

In general the equations of state can be divided into two categories: analytical EOS and tabular EOS.

In LS-DYNA [5] the equations of state for metals are in general partitioned into two terms: a cold pressure and a thermal pressure (see Eq. 7). The first terms $P_C(\mu)$ is a function of the density and is hypothetically evaluated along a 0 K isotherm, while the second one $P_T(\mu,E)$ depends on both the density and the internal energy per unit volume of the system

$$P(\mu, E) = P_C(\mu) + P_T(\mu, E) \tag{7}$$

where $\mu = \rho/\rho_0 - 1$. It is clear that for a compressed material $\mu > 0$, for an expanded material $\mu < 0$ and if no load is applied to the material $\mu = 0$. More precisely the EOS implemented have the general form

$$P(\mu, E) = A(\mu) + B(\mu)E \tag{8}$$

so the pressure is a general function of the density but linearly depending by the energy per unit volume.

An example of analytical equation of state implemented in LS-DYNA is the Mie-Grüneisen EOS [5], that is able to describe the behaviour of a solid material. It defines the pressure for compressed materials as

$$P = \frac{\rho_0 C_0^2 \mu \left[1 + \left(1 - \frac{\gamma_0}{2}\right)\mu - \frac{a}{2}\mu^2\right]}{\left[1 - (S_1 - 1)\mu - S_2\frac{\mu^2}{1+\mu} - S_3\frac{\mu^3}{(1+\mu)^2}\right]^2} + (\gamma_0 + a\mu)E \tag{9}$$

and for expanded materials as

$$P = \rho_0 C_0^2 \mu + (\gamma_0 + a\mu)E \tag{10}$$

where ρ_0 is the initial density, C_0 the sound speed, γ_0 the Grüneisen parameter that defines the effect on the atoms vibration consequent to the change in energy (or temperature), S_i are the coefficients of the polynomial that defines the shockwave velocity as function of particle velocity and a is the first volume correction to γ_0.

Another possible form of an equation of state is the tabular form. For the construction of the tabular EOS several codes are developed, an example is EOSPro code [24]. In this particular case the EOS includes solid and fluid phases and the dependent variable (i.e. pressure or internal energy) is defined as function of independent variables (temperature and density), as shown in Fig. 2.

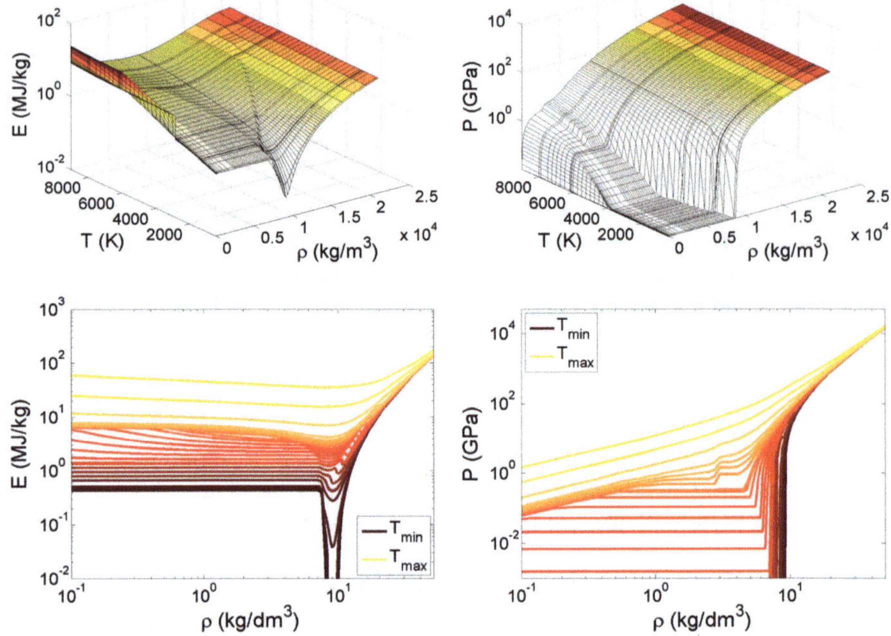

Fig. 2 Tabular EOS of copper: internal energy and pressure in function of temperature and density [22]

5 Case Study

In the present work the case study is the impact of 8 bunches at 7 TeV on a cylindrical copper bar that is 1 m long and 5 mm in radius. The case study represents an abnormal situation in which the beam impacts directly against the solid target perpendicular to the base of the cylinder. A non reflection boundary condition is applied at the external radius to model the presence of other material.

The energy deposition on the hit material is calculated by the Fluka team at CERN using the FLUKA code on the same geometry for a solid copper target. FLUKA provides the distribution on a 2D axisymmetric model counting 1,000 elements both in radial and axial direction for the interaction with a single proton. In Fig. 3 the specific energy deposition due to the impact with one bunch is plotted in function of the longitudinal coordinate at different radial positions and in function of the longitudinal coordinate at different transverse positions. As mentioned before, in each bunch there are 1.15×10^{11} protons and each bunch is 0.5 ns long and the time between two consecutive bunches is 25 ns so the total duration of the impact is about 200 ns.

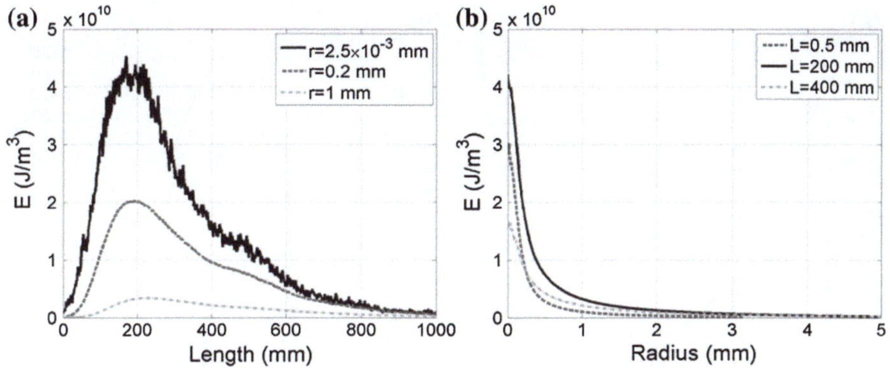

Fig. 3 Energy deposition: **a** longitudinal distribution for different radii; **b** radial distribution for different longitudinal positions

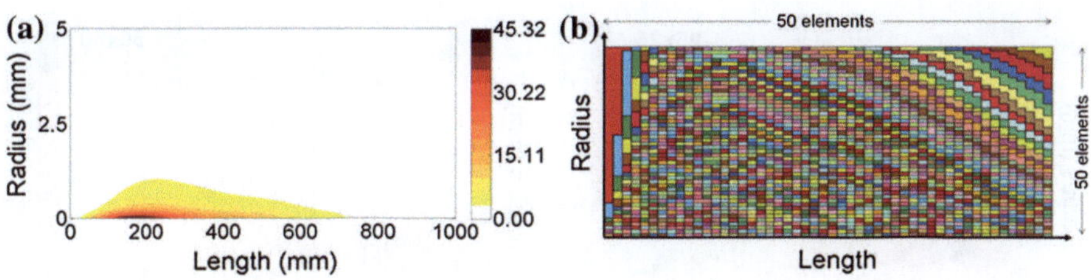

Fig. 4 FLUKA energy distribution (GJ/m^3) for a single bunch (**a**); numerical model (**b**)

5.1 FEM Model

The model used in this simulation is a 2D axisymmetric model with the same geometry on which the FLUKA calculation is performed. It is chosen a 2D axisymmetric model because in LS-DYNA the complexity to realize 2D or 3D models is the same. Besides, in this way the comparison with the results obtained in [12] whit the hydrocode BIG-2 is easier.

The energy distribution per unit bunch calculated by FLUKA code and deposited on the component is shown in Fig. 4.

Four types of mesh are analyzed in order to investigate the influence of the spatial discretization on the results. The comparison is performed on the starting case in which the material is considered to remain solid, so the Mie-Grüneisen EOS is used for the pressure, energy and density calculations, and the energy deposition is approximated as an initial condition. The models that are simulated present the following elements densities: 25 × 25, 50 × 50, 100 × 100 and 200 × 200 elements in the radial and longitudinal directions. In Fig. 5 the comparison is performed in terms of pressure history variable for two elements and Figs. 6 and 7, respectively, show the results in terms of pressure and density at the

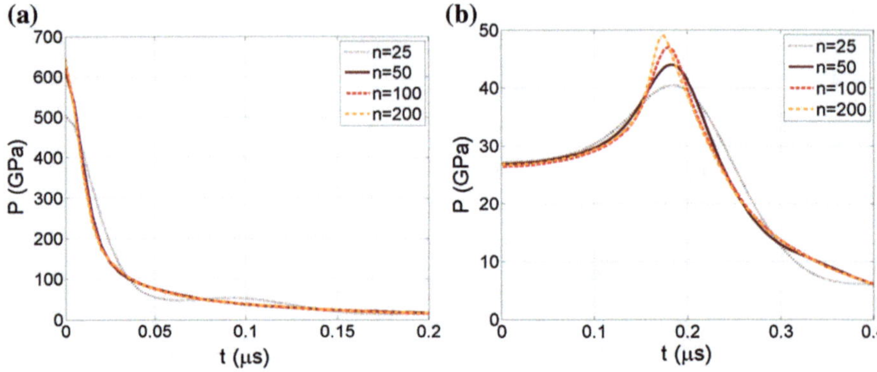

Fig. 5 Comparison in terms of pressure (GPa) in two different points varying the mesh densities: **a** $r = 0$ mm, $l = 200$ mm; **b** $r = 1.2$ mm, $l = 400$ mm

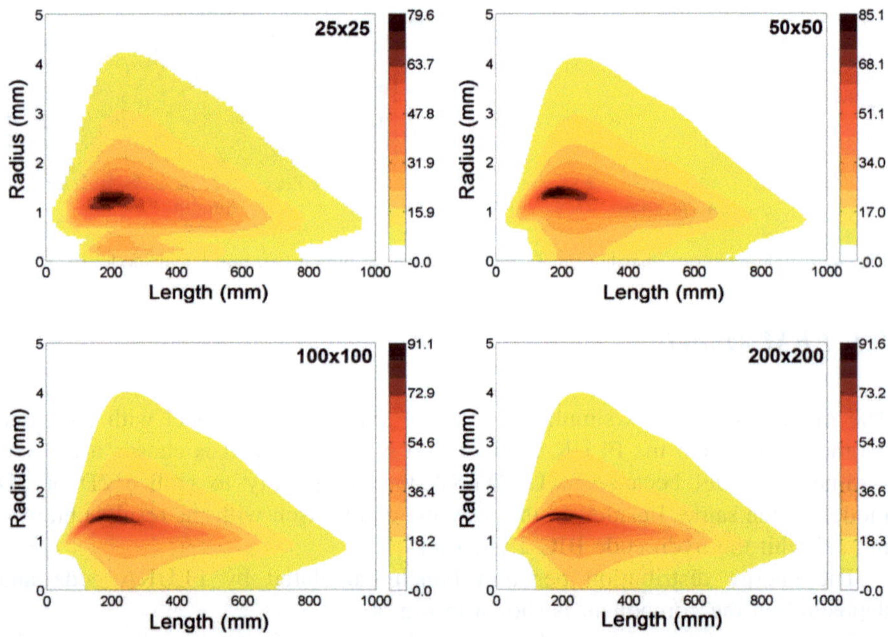

Fig. 6 Comparison of the result obtained with different mesh densities in terms of pressure (GPa) at the end of deposition (about 200 ns)

end of deposition. Obviously, there are some differences in the solutions due mainly to the FEM accuracy and the precision in the allocation of the FLUKA distribution in the numerical model.

The material model used in the simulations is the Johnson-Cook material model [5], and the input parameters are summarized in the Table 1.

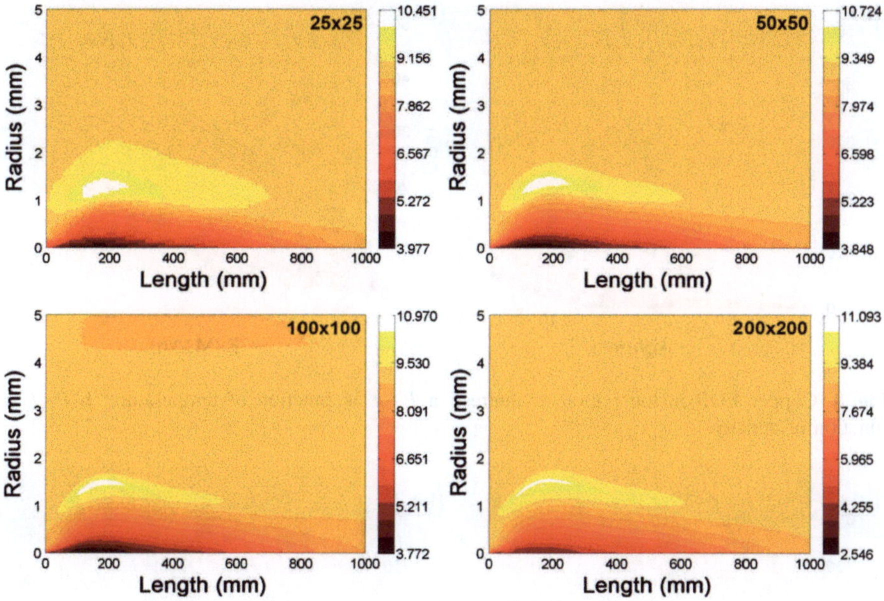

Fig. 7 Comparison of the result obtained with different mesh densities in terms of density (kg/dm^3) at the end of deposition (about 200 ns)

Table 1 J–C parameters

Parameter	Value	Unit	Parameter	Value	Unit
ρ_0	8,937.5	kg/m^3	T_{melt}	1,356	K
G	4.6×10^{10}	Pa	T_{room}	300	K
E	1.11×10^{11}	Pa	C_p	383	J/(kg K)
v	0.3406	–	$P_{\text{cut-off}}$	1.2×10^9	Pa
A	9.00×10^7	Pa	D_1	0.54	–
B	2.92×10^8	Pa	D_{22}	4.89	–
n	0.31	–	D_{33}	−3.03	–
C	0.025	–	D_{44}	0.014	–
$\dot{\varepsilon}_0$	1.00	s^{-1}	D_{55}	1.15	–
m	1.09	–			

In LS-DYNA code there isn't the possibility to use a tabular equation of state in the same form of that is calculated in [25]. To solve this problem, the code offers the possibility to implement an user-defined EOS. In this work a standard polynomial EOS is used to fit the tabular data to evaluate a simplified approach

$$P = C_0 + C_1\mu + C_2\mu^2 + C_3\mu^3 + \left(C_4 + C_5\mu + C_6\mu^2\right)E \tag{11}$$

The multi-phase EOS for copper is well described in [25] and represented in Figs. 2 and 8a. In order to obtain the correct interpolation with the Eq. (11) a new plot is created drawing the function $P = P(E)$ for each density (see Fig. 8.b).

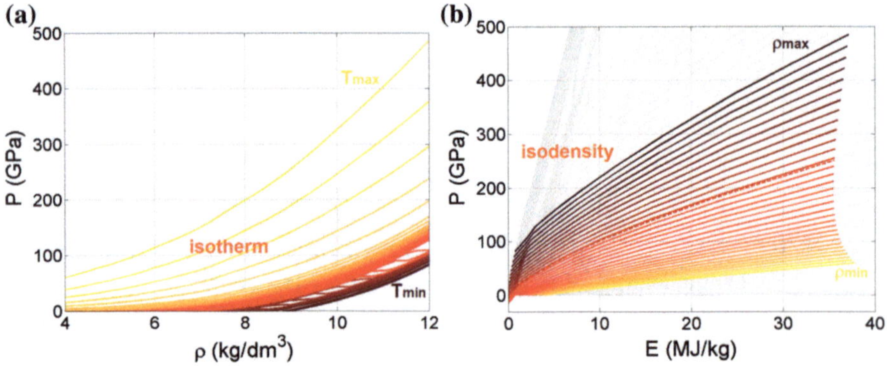

Fig. 8 Copper EOS in the region of interest: **a** P–ρ as function of temperature; **b** P–E as function of density

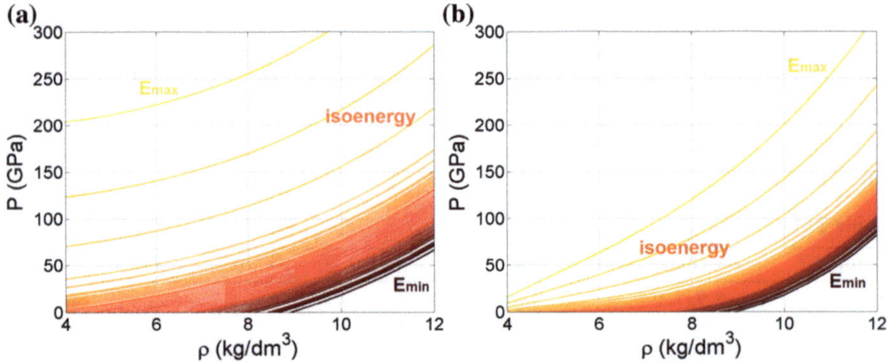

Fig. 9 Results of polynomial interpolation: **a** solid phase; **b** entire tabular data

The relation between P and E for a fixed ρ could be approximate with a generic function in E. Since the standard formulation (see Eq. 11) is linear with the energy the approximation will be more or less coarse depending on the energy range considered.

Below there is the description of the interpolation methods used to get the polynomial coefficients to set as numerical simulation inputs.

The first set of simulations is performed assigning to all elements the parameters extracted from the interpolation of the solid phase data (see Fig. 9a) and presented in the Table 2. In this case the region of interest, for what concerns the interpolation, is limited between 9,000 and 12,000 kg/m^3 in density and 5 and 1,360 K in temperature. As it could be expected interpolation is good in high density and low energy range which includes the solid matter behaviour, while in the other parts it is very coarse. Obviously, the consequence in the simulations results will be the pressure overestimation in the elements that will reach low density and/or high energy values.

Table 2 EOS parameters (solid phase)

C_0 (Pa)	C_1 (Pa)	C_2 (Pa)	C_3 (Pa)	C_4 (–)	C_5 (–)	C_6 (–)
2.079×10^7	1.377×10^{11}	1.256×10^{11}	1.454×10^{11}	1.940	0.585	1.125

Table 3 EOS parameters (entire tabular data)

C_0 (Pa)	C_1 (Pa)	C_2 (Pa)	C_3 (Pa)	C_4 (–)	C_5 (–)	C_6 (–)
-2.176×10^8	1.113×10^{11}	3.019×10^{11}	2.088×10^{11}	1.076	1.667	–0.039

In order to check the method of procedure for the solid coefficients the comparison with the Mie-Grüneisen equation of state is performed and the results obtained with this two formulations are equivalent.

Then a second set of simulations is performed using the parameters extracted considering the whole region of interest. In this case the density is between 4,000 and 12,000 kg/m^3 and the temperature is between 5 and 27,000 K. The fit is carried out in two phases. First, the curve P–μ at 5 K is interpolated to obtain the coefficients of the cold curve $P_C(\mu)$ (see expression 11). Then the coefficients for the $P_T(\mu,E)$ are get fitting with a second-order polynomial the slopes calculated from the linear interpolation in energy on the tabular data. The results in term of pressure in function of density varying the energy is shown in Fig. 9b. Comparing this with the Fig. 8a, it is clear that the approximation is quite good except for the phase transition. The parameters are summarized in the Table 3.

In Figs. 10 and 11 there are the results obtained with the two different sets of parameters at two different time steps. The simulations are performed on the 50 × 50 elements mesh under the assumption that the energy deposition can be approximated as an initial condition.

As mentioned before, the use of the equation of state obtained fitting the solid phase data respect to the equation of state related to the entire tabular data produces a great pressure overestimation. In fact, for example, for a fixed time the pressure obtained in the first case is approximately twice the pressure in the second one. Besides, you can see that for a fixed time the pressure and density profiles get deeper in the component in the radial direction with the solid interpolation: this is due to the difference of the shockwave speed, consequent to the variation in density and bulk modulus.

5.2 Energy Deposition Method

The energy deposition is simulated by using three different methods:

- the energy stored in 8 bunches is applied to the component as an initial condition;
- the time-energy profile is a 204 ns ramp (constant power);
- the time-energy profile is the sequence of the 8 bunches: each of them is 0.5 ns constant power and 25 ns void.

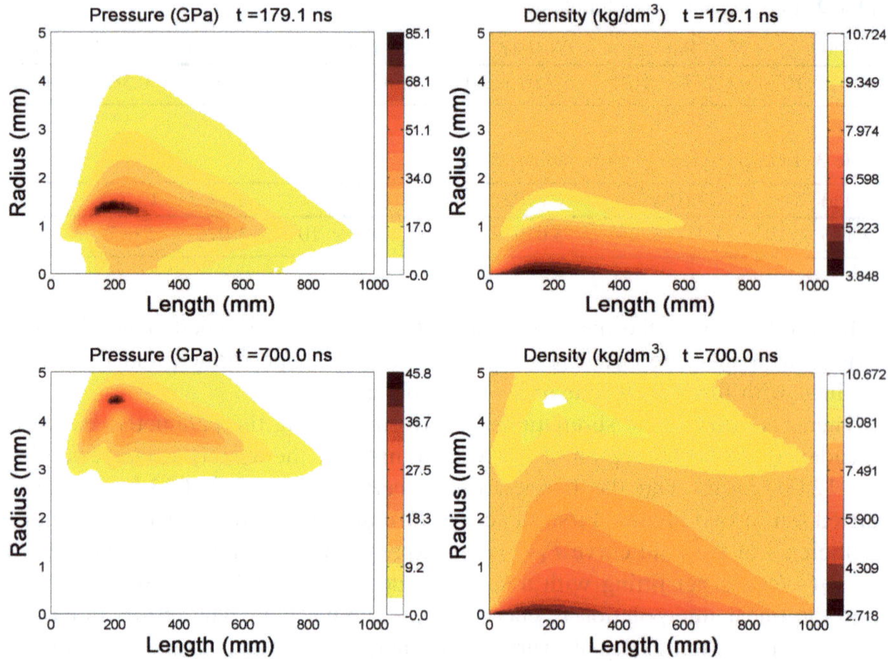

Fig. 10 Pressure and density evolutions obtained for the simulations performed using for the EOS definition the parameters extracted fitting the solid phase tabular data

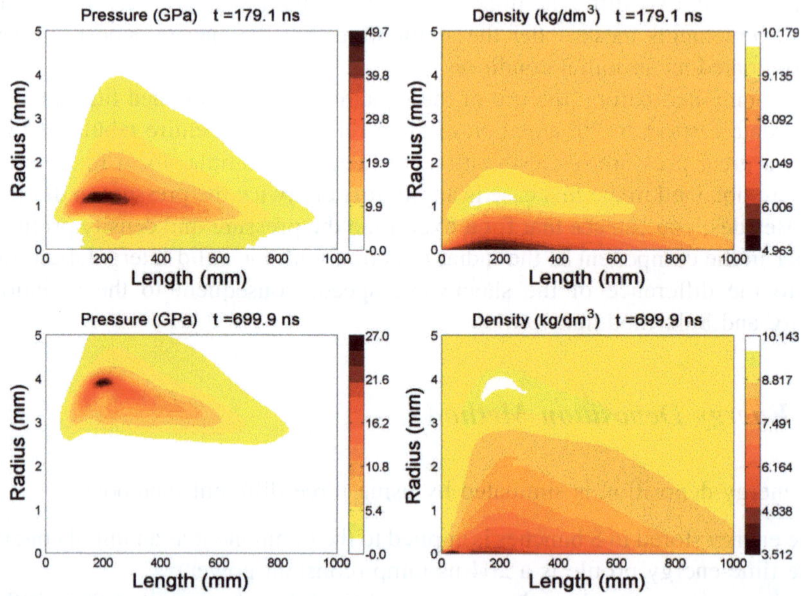

Fig. 11 Pressure and density evolutions obtained for the simulations performed using for the EOS definition the parameters extracted fitting the entire tabular data

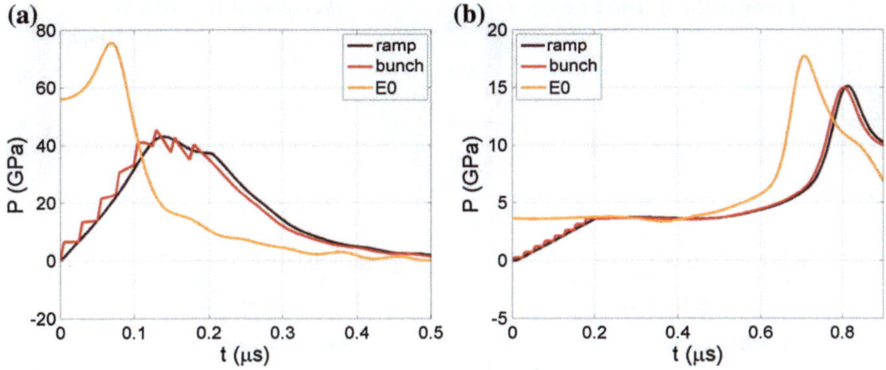

Fig. 12 Comparison of the pressure time history in two elements for the different load cases: **a** r = 0.55 mm, l = 250 mm; **b** r = 3.85 mm, l = 250 mm

The way to reproduce an energy deposition in LS-DYNA is defining for each element (or for all the elements in which the energy level to be deposited falls within a certain range) a specific equation of state with the corresponding deposition time-history.

In Fig. 12 the comparison between the different deposition methods is focalized on the pressure time history in two different elements. The comparison is carried out for a 50 × 50 elements mesh with the EOS coefficients obtained from the fit of the entire tabular equation of state. In Fig. 13 there is the comparison in term of pressure at three different time steps for the extreme cases (initial condition and bunches profile). As it can be seen, if the energy is transferred to the material as an initial condition it produces an overestimation of the pressure, while if the correct bunches profile is used the pressure level reached is lower because the material has the time (25 ns of void between two successive bunches) to expand and reduce its pressure.

From the Fig. 12 it is clear that the ramp and the bunches profiles generally produce the same results except for the deposition duration (about 200 ns). In fact, during the deposition, there are some diversities due to the different deposition methods, but in both cases the maximum energy level reached is the same and so the material condition after the end of deposition is quite similar.

6 Results and Discussion

In this section the attention is focused on the results of the numerical simulation performed with 50 × 50 elements mesh, the EOS coefficients obtained fitting the entire tabular EOS data with the polynomial formulation and the bunches deposition profile. This configuration appears to be a model sufficiently adequate and

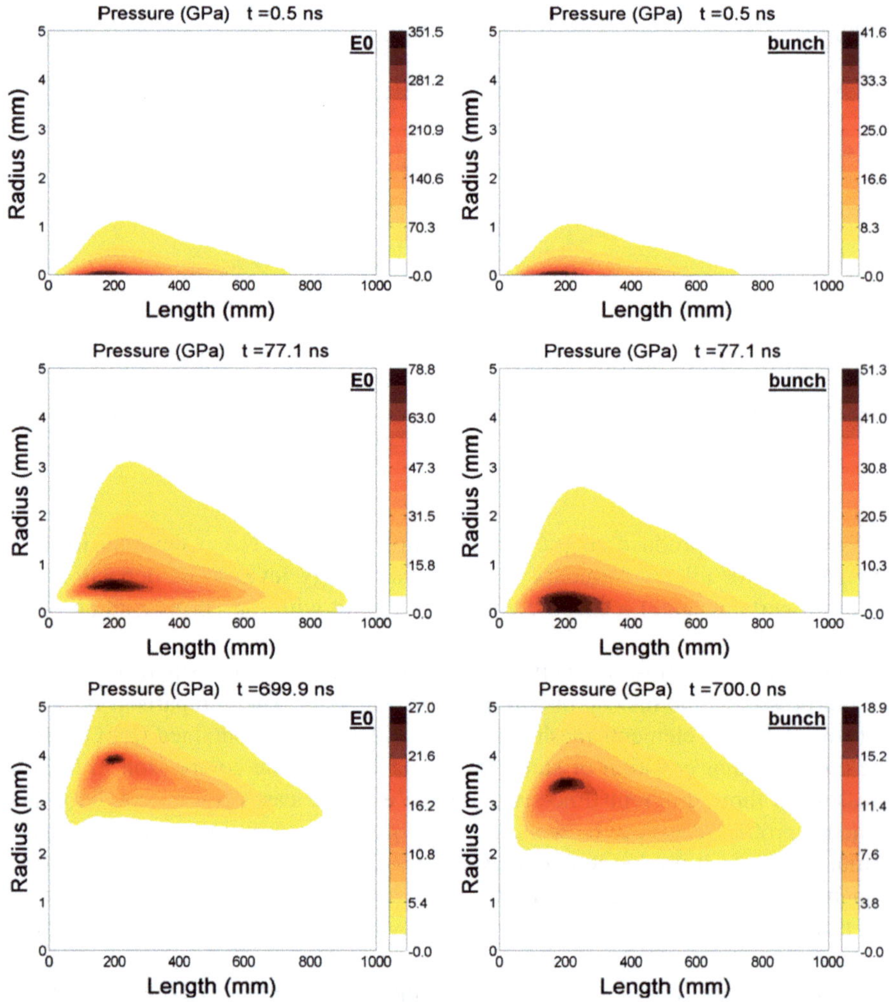

Fig. 13 Comparison in terms of pressure (GPa) at three different time steps in case of energy deposition as initial condition (on the *left*) and with the bunches profile (on the *right*)

representative of the evolution of the phenomenon. Figure 14 shows the results in terms of pressure, density and Von Mises stress at different time steps.

Regarding the pressure, the energy deposition delivered after the impact of 8 bunches produces a very high level of pressure: a maximum value of 46.9 GPa at the end of deposition phase while the maximum value reached at a radius of 5 mm is about 17.2 GPa (after 1 μs).

The impact generates the outgoing of shockwaves that travelling the component produces a density reduction in the central part of the cylinder. It is highlighted by Von Mises stress distribution that over than 50 % of the target is molten (in this

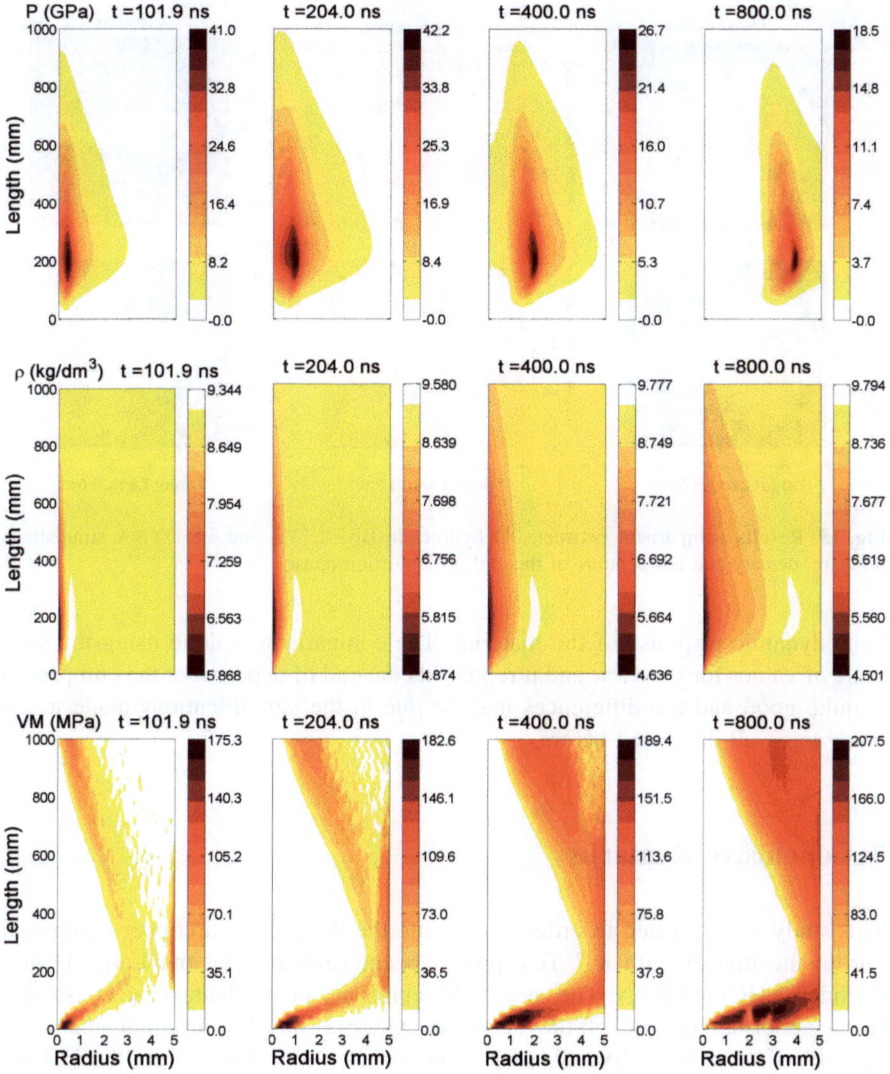

Fig. 14 Numerical results: pressure, density and Von Mises stress at four different time steps during the phenomenon evolution

part the temperature exceeds the melting value so the stress flow calculation goes to zero). The remaining portion of the material is heavily plastically deformed consequent to the stress wave propagation.

In order to verify the likelihood of the solution it is compared (see Fig. 15) with the results obtained in [12], in which the simulations are performed via a 2D hydrodynamic code BIG-2, which is developed to solve this kind of problems. Besides, in [12] a more accurate tabular equation of state is used to describe the

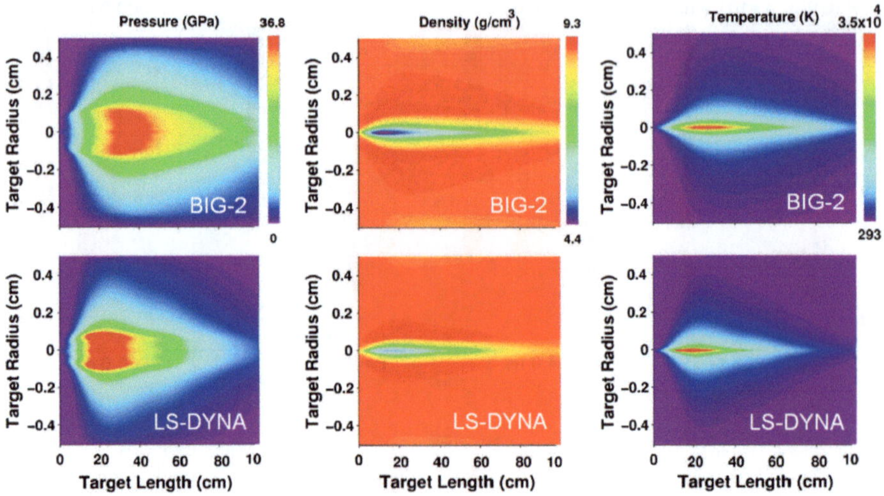

Fig. 15 Results comparison between 2D hydrocode BIG-2 [12] and LS-DYNA simulations: pressure, density and temperature at the end of deposition phase

hydrodynamic response of the material. The comparison is done using the same scale of values for the view and it represents the end of deposition: the comparison is quite good and the differences may be due to the simplifications made in this work and well discussed before.

7 Conclusive Remarks

This study is performed in order to estimate the damage on a copper component due to the impact with a 7 TeV proton beam generated by the Large Hadron Collider (LHC) at CERN. The case study represents an accidental case consequent to an abnormal release of the beam, in which 8 bunches irradiate the target directly. The energy delivered on the component is calculated using the FLUKA code and then used as input in the numerical simulation, that are carried out via the FEM code LS-DYNA.

In this kind of problems both the hydrostatic and the deviatoric components are involved, but in different regions of the hit component. The hydrostatic behaviour of the target is described using a polynomial equation of state obtained fitting a multi-phase EOS, while the elasto-plasticity response is treated using the Johnson-Cook material model.

Different numerical models are realized trying to obtain the simplest model able to correctly describe the material response without affecting the goodness of the results. The most suitable, in terms of accuracy of the results and computational speed, is found to be the model with 50 $\times$ 50 elements mesh, the EOS coefficients obtained fitting the entire tabular EOS data with the polynomial formulation and the bunches deposition profile. This configuration appears to be a model sufficiently adequate and representative of the evolution of the phenomenon.

The material closed to the beam heated region reaches high value of pressure (46.9 GPa after the irradiation of 8 bunches) and can melt or vaporize. In this part the simulation indicates a reduction in density (until 4,500 kg/m^3) due to the radial shockwaves travel. When the pressure shockwave reaches the radius of 5 mm about the 50 % of the component is damaged. The remaining portion of the material is subjected to high stress, plastic deformation, strain-rate and temperature even it is still solid.

Acknowledgments This work was performed within the WP 8 "Collmat" of the FP7 European Project EUCARD. The financial support of the European Commission by means of the EUCARD project, the Fluka Team and Gerald Kerley (Kerley Technical Services) are gratefully acknowledged.

References

1. Wiedemann, H.: Particle accelerator physics I. Springer, Berlin (1993)
2. Petterson, T.S., Lefèvre, P.: The large hadron collider: conceptual design. CERN Desktop Publishing Service, Geneve (1995)
3. LHC Design Report, vol. I, The LHC Main Ring. CERN Editorial Board (2004)
4. LHC Design Report, vol. III, The LHC Injector Chain. CERN Editorial Board (2004)
5. Gladman, B et al.: LS-DYNA® Keyword User's Manual—Volume I—Version 971. LSTC (2007)
6. Fasso, A. et al.: FLUKA: A Multi-Particle Transport Code (2005)
7. Dallocchio, A.: Study of termo-mechanical effects induced in solids by high energy particle beam: analytical and numerical methods. Ph.D Thesis, Politecnico di Torino (2008)
8. Bennett, J.R.J., Booth, C.N., Brownsword R.A et al.: LS-DYNA calculation of shock in solid. Nucl. Phys. B (Proc. Suppl.) **155**, 293–294 (2006)
9. Bennett, J.R.J., Skoro, G.P., Booth, C., et al.: Thermal shock measurements and modeling for solid high-power targets at high temperatures. Nucl. Mater. **377**, 285–289 (2008)
10. Mokhov, N. http://www-ap.fnal.gov/MARS
11. Tahir, N.A., Goddard, B., Kain, V et al.: Impact of 7-TeV/c large hadron collider proton beam on a copper target. Appl. Phys. 97 (2005)
12. Tahir, N.A., Shutov, A., Lomonosov IV et al.: Thermo-mechanical effects induced by beam impact on the LHC Phase II collimators: preliminary analysis using hydrodynamic approach
13. Le Blanc G, Petit J, Chanal P et al.: Modelling the dynamic magneto-thermomechanical behavior of materials using a multi-phase EOS. In: 7th European LS-DYNA Conference (2008)
14. Johnson, J.R., Cook, W.H.: A constitutive model and data for metals subjected to large strains, high strain rates and high temperatures. In: 7th International Symposium on Ballistics, pp. 541–547 (1983)
15. Cowper, G.R., Symonds, P.S.: Strain hardening and strain rate effects in the impact loading of cantilever beam. Brown University Division of Application of Mechanical Report 28 (1952)
16. Zerilli, F.J., Armstrong, R.W.: Dislocation-mechanics-based constitutive relations for material dynamics calculations. J. Appl. Phys, Vol. 61 (1987)
17. Peroni, M., Peroni, L., Dallocchio, A.: Thermo-mechanical model identification of strengthened copper with an inverse method. 9th International DYMAT Conference (2009)

18. Peroni, M.: Experimental methods for material characterization at high strain-rate: analytical and numerical improvements. Ph.D thesis, Politecnico di Torino (2008)
19. Peroni, L., Scapin, M., Peroni, M.: Identification of strain-rate and thermal sensitive material model with an inverse method. In: 14th International Conference on Experimental Mechanics (2010)
20. Steinberg, D.J., Cochran, S.G., Guinan, M.W.: A constitutive model for metals applicable at high-strain rate. J. Appl. Phys. **51**, 1498 (1980)
21. Steinberg, D.J., Lund, C.M.: A constitutive model for strain rates from 10^{-4} to 10^{6} s^{-1}. Journal de physique, Symposium **C3**(49), 433–440 (1988)
22. Macdougall, D.: Determination of the plastic work converted to heat using radiometry. Exp. Mech. **40**, 298–306 (2000)
23. Hodowany, J., Ravichandram, G., Rosakis, A.J., et al.: Partition of plastic work into heat and stored energy in metals. Exp. Mech. **40**, 113–123 (2000)
24. Kerley, G.I. http://www.kerleytechnical.com/eospro.htm
25. Kerley, G.I.: Equation of State for Copper and Lead. Kerley Technical Services Report KTS02-1 (2002)

Modeling of High Velocity Impact in Sandwich Beams with FGM Core

J. Zamani, E. Etemadi, K. Hosseini Safari and A. Afaghi Khatibi

Abstract In this work, the impact behavior of sandwich beams with asymmetrical and symmetrical Functionally Graded (FG) cores are investigated using finite element method. The ballistic limit and residual velocity of several sandwich beams with a FG core are studied. The ratios of residual kinetic energy as well as perforation energy to initial kinetic energy are calculated. Furthermore, the effects of failure strain, tangent modulus and thickness of face sheets on the behavior of sandwich beams are discussed. Additionally, the effects of initial velocity of the projectile on the damage extension of sandwich beam are thoroughly investigated. Results indicate that the strength of beams with asymmetrical FG core under high velocity impact is more than that of beams with symmetrical one.

Keywords Functionally graded materials · Sandwich beams · High velocity impact · Finite element · Ballistic limit

1 Introduction

Sandwich structures usually are made of two thin and stiff face sheets that are separated with a thick and low density core. They have been extensively used in aerospace and automobile applications and the understanding of their behavior

J. Zamani (✉) · E. Etemadi · K. H. Safari
Faculty of Mechanical Engineering, K. N. Toosi University of Technology,
P.O. Box 19395-1999, Tehran, Iran
e-mail: zamani@kntu.ac.ir

E. Etemadi
e-mail: Etemadi@dena.kntu.ac.ir

K. H. Safari
e-mail: safari@dena.kntu.ac.ir

A. A. Khatibi
American University of Sharjah, P.O. Box 26666, Sharjah, UAE

N. Bonora and E. Brown (eds.), *Numerical Modeling of Materials Under Extreme Conditions*, Advanced Structured Materials 35, DOI: 10.1007/978-3-642-54258-9_5,

under impact conditions is extremely important for the further development of these materials [1].

For a sandwich structure subjected to low velocity impact loading by a projectile, the main parameters to be investigated through developing of dynamic models are (1) projectile motion, (2) local and global deformation, (3) contact area and (4) the behavior of sandwich structure. Several types of mathematical models are used to study the impact of a structure. The classical method decouples the local and global responses and ignores any interaction between the two. The first order shear deformation theory (FSDT) and higher-order shear deformation theories (HSDT) and Bernoulli-Euler beam theory and Timoshenko beam theory [2, 3] do not consider the flexibility of the core in transverse direction. In addition, the interaction between face sheets and soft and flexible core is also neglected. Most recently, the higher-order sandwich plate theory (HSAPT) [4] and improved higher-order sandwich plate theory (IHSAPT) [5] has been introduced.

Contact force is an important parameter in low velocity impact problems and can be characterized accurately by dynamic models. Spring-mass models have accurate answers for small size specimens and quasi-static responses [6]. Anderson [7] added damper to a spring-mass mechanism in order to model the damage in a composite sandwich structure. Calculating the maximum contact force is also possible when the structure behaves quasi-statically using Energy balance approach. In this case, it is assumed that when the contact force reaches its maximum, the sum of the strain energy in the structure and the energy required for indentation is equal to the initial kinetic energy of the projectile. It has been shown that when the deformation of plate is negligible and only the dynamic behavior of plate is important, the energy balance approach is useful [8]. Olsson [9] developed an approximate solution for infinite composite plates by defining an inelasticity parameter which depends on the contact stiffness, ratio of mass to contact area, bending stiffness of the structure, and velocity as well as mass of the projectile. Evaluations from this model were in good agreement with experimental data when it was used in an impact test with a projectile of small mass and high velocity [6].

Although several experimental work have been conducted to investigate the various types of impact in sandwich structures [10–13] but usually it is not possible to change all effective parameters in these experiments and mainly these experiments are time-consuming and costly. Along with experimental methods, numerical analyses such as finite element method are used extensively to model the impact behavior of more complicated loading cases and geometries in sandwich composite structures [14–17].

One of the problems of sandwich structures is their relatively low resistance to impact damage due to the thin outer composite skins. When a composite skin is subjected to impact by a projectile, many of the fracture processes, with their associated energy absorbing capacities, can occur [18, 19]. The major damage mechanisms such as matrix cracking, debonding and fiber failure may appear individually. However, their interaction with the effects of fiber type and lay-up, distribution and geometrical nature, the type and state of the matrix-fiber bonding as well as environmental effects could result in catastrophic failure [20].

For relatively low velocity projectiles, a sandwich beam may respond by bending and no damage will occur if the energy of the projectile can be accommodated by the elastic strain energy of the beam. At higher impact velocities, a critical condition is reached when a local contact stress exceeds the local strength, which may be either a laminate bending strength, core compression strength or interface delamination strength [21, 22].

Other important problems in sandwich structures are the damage due to discontinuity of mechanical properties at the boundary of face sheets and core. When these structures are exposed to impact loading, for example, shear stresses can cause the debonding of face sheets from core. Wu and Sun [23] and Shipsha et al. [24] investigated different failure modes of sandwich beams and suggested that these structures can be upgraded using new materials in face sheets and cores.

Utilizing a Functionally Graded (FG) core in sandwich panels is increasing because of their capabilities in reducing thermal- and residual-stresses induced between the face sheets and core materials in comparison to conventional sandwich panels. Venkataraman and Sankar [25] indicated that a graded core expressively could reduce the face sheet-core interfacial shear stresses while Nakamura and Wang [26] demonstrated the possibility of reducing impact damage in FGMs using finite element method. A series of test cases has been developed by Nettles and Hodge [27] to investigate the critical parameters associated with the effective modeling of composites and sandwich beams. In these tests different projectiles, such as hard and soft ones together with a gas gun can be used to achieve the required impact velocity [28, 29]. In addition, numerical methods such as finite element [30, 31], Smooth Particle Hydrodynamics (SPH) [32] and discrete element [33] were used to study the impact behavior of these beams.

Some works have been done on the impact behavior of sandwich panels with functionally graded cores. Anderson [34] used analytical three-dimensional elasticity solution to define the contact between a rigid spherical projectile and a sandwich panel while the ratio of thicknesses of face sheet to core was 0.1. Apetre et al. [35] used a polynomial variation of Young's modulus of core and Galerkian method to describe the impact of a sandwich beam having small ratio of face sheet to core thicknesses (0.015) with a cylindrical projectile. They showed that FG core can significantly reduce the shear and normal strains. Kashtalyan and Menshykova [36] with three-dimensional elasticity theory showed that the use of a graded core instead of a conventional homogeneous one eliminates discontinuity of the in-plane normal and shear stresses across the face sheet–core interfaces.

Kirugulige et al. [37] compared the fracture behavior of sandwich structures with a FG core and conventional sandwich structures using experimental as well as two dimensional FEM. Etemadi et al. [38] simulate sandwich panels with a functionally graded core under low velocity impact using 3D finite element.

A new finite element specifically developed for thermo-elastic analysis of sandwich panel with FGM core was proposed by [39]. Zhu and Sankar [40] studied response of the sandwich panel to the transient heat conduction. Again, variation of the core Young's modulus was represented by a polynomial in the thickness co-ordinate. An experimental investigation of failure of piece-wise functionally

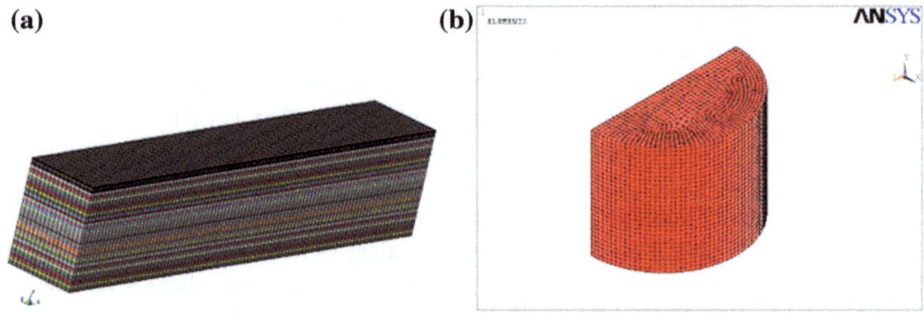

Fig. 1 FE modeling one fourth of **a** Sandwich beam with FG core, and **b** Projectile

graded of sandwich beams subjected to three-point bending was carried out by Avila [10].

In order to utilize the full potential of new sandwich beams with a FG core, it is necessary to investigate their impact characteristics using cost effective methods. In this work, the high velocity impact behavior of a sandwich beam with FG core is studied using 3D finite element method. The effects of symmetrical and asymmetrical FG core, initial velocity of the projectile as well as geometrical parameters on the impact behavior of sandwich beams are investigated.

2 Modeling of the Problem

In this study, the FG core was divided to 41 small volumes and Young's modulus was then assigned to these volumes individually.

The FE model consists of a steel cylinder projectile and a sandwich beam having two thin face sheets with a functionally graded core. The mass of the projectile was $m = 15.7$ Kg with the radius of $r = 0.01$ m. For the sandwich beam, the thickness of face sheets was $h_f = 0.3$ mm with Young's modulus of $E_f = 500$ GPa and Poisson's ratio of $v_f = 0.25$. The length and width of the beam are 0.2 and 0.03 m. The variation of Young's modulus in the thickness of FG core ($2\text{ h} = 0.02$ m) assumed to be linear symmetric E_{Sym} and/or linear asymmetric E_{Asym} about the mid plane and is given by:

$$E_{\text{Sym}} = E^0\left(\frac{E^h - E^0}{E^0}\frac{|z|}{h} + 1\right) \quad \text{when} \quad -h \le Z \le h \tag{1a}$$

$$E_{\text{Asym}} = E^{-h}\left(\frac{E^h - E^{-h}}{E^{-h}}\frac{|z|}{h} + \frac{E^{-h} + E^h}{2E^{-h}}\right) \quad \text{when} \quad -h \le Z \le h \tag{1b}$$

where Z is the coordinate along the thickness of the beam, E^0, E^h and E^{-h} refers to the Young's modulus of FG core at the midplane, top and bottom interfaces, respectively.

Fig. 2 Contact states during high velocity impact for one forth of a cylindrical projectile and sandwich beam

Figure 1 shows the variations of Young's modulus through the thickness for a functionally graded core with Homogenous (HC), Symmetrical (SC) and Asymmetrical (AC) conditions. It should be noted that the Poisson's ratio for all FG cores was assumed to be constant and equal to $v = 0.35$.

SOLID elements with 8 nodes were used in FE simulations for the projectile as well as the target. The projectile was considered as rigid while the sandwich beam treated as an MAT-PLASTIC-KINEMATIC. Considering constant mechanical properties in length (x-direction) and width (y-direction) of the beam and geometrically symmetric conditions, only one forth of the projectile and target were modeled. The number of elements for the target was 100 (length) $\times$ 15 (width) $\times$ 43 (height) $=$ 73,500 (indicated as mesh 1) while the projectile in all cases had 24,695 elements (See Fig. 2).

It should be noted that under high velocity impact, damage and failure may happen in structures and therefore the material properties should be defined as elastic-plastic. To this end, for sandwich beams in the current model, material properties of MAT-PLASTIC-KINEMATIC was selected which is defined based on a bilinear stress-strain diagram. In this material model, several parameters should be defined. Tangent modulus (E_t) is one of the input parameters. The simplest method to describe the tangent modulus is the slope of stress-strain diagram in plastic region with the assumption that the ultimate tensile stress happens at the failure strain. Functionally graded materials exhibit work-hardening behavior after reaching the yield stress [41]. For a functionally graded material (FGM) phase, relation between its tangent and Young's modulus is given by [42]:

$$E_t = \frac{E}{1 + \frac{E}{H}} \tag{2}$$

where E and H are Young's modulus and work-hardening ratio of each phase, respectively. The other parameter which is necessary in elastic-plastic behavior is the failure strain. To design structures under dynamic loading, the relation between stability of the material and applied strain rate should be considered. The yield stress of the material could then be calculated. The type of contact that has been used in this study was Eroding Surface to Surface (ESTS) contact. This type of contact can prevent from decreasing time steps due to weakened elements in the contact region which occurs in large deformations.

The tangent modulus was considered as 2 GPa for FG core and 1 GPa for the face sheet material. Failure strain for the layers of FG core as well as face sheets were 0.1 and 0.02, respectively. Moreover for all of the cases studied in this work, the yield stress corresponding to E = 50 GPa of FG core was 96 MPa while for functionally graded core with E = 500 GPa it was 180 MPa. The yield stress of FG core for different values of Young's modulus varies between two abovementioned values, i.e. 96 and 180 MPa.

In high velocity impacts, the important parameters are velocities of a given projectile at the beginning and end of contact. Also, it's important to know the ballistic limit, i.e. minimum impact velocity that will result in complete penetration [43, 44]. Additionally, residual velocity is the velocity of projectile when exits completely from the target and can be considered as one of important parameters. It's clear that for the ballistic limit, the residual velocity is equals to zero. The relation between initial velocity (V_i) and residual velocity (V_r) is given by:

$$\mathrm{U_P} = \frac{1}{2} M\left(Vi^2 - Vr^2\right) \tag{3}$$

where U_P is the perforation energy and M is the mass of projectile.

Figure 2 shows the Contact states during high velocity impact for one forth of a cylindrical projectile and sandwich beam. The instant velocity history for a sandwich beam with an asymmetrical FG core under impact loading with a projectile is sensitive to mesh numbers. It's clear that increasing the number of elements in the height direction of the beam is more effective than increasing the number of elements in its x and y directions. Considering the results from this study, it can be concluded that in high velocity impact simulation, the mesh density is an important parameter.

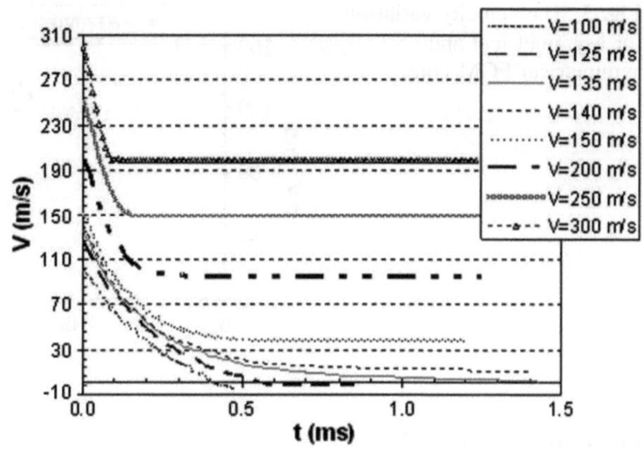

Fig. 3 Instant velocity history of a sandwich beam with an asymmetrical FG core

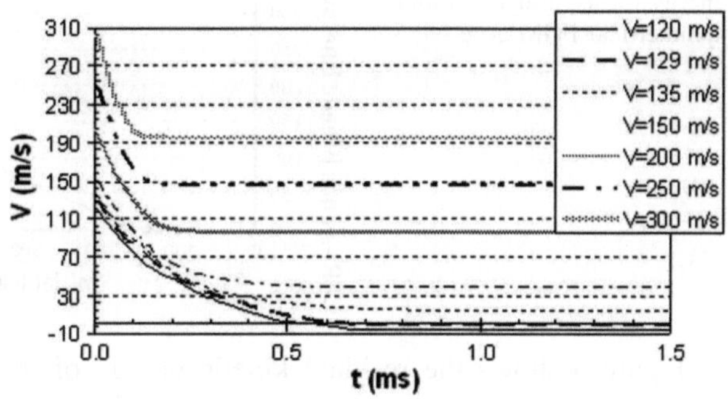

Fig. 4 Instant velocity history of a sandwich beam with a symmetrical FG core

3 Ballistic Impact

In Figs. 3 and 4 the effects of initial velocity (V_i) on the instant velocity history of sandwich beams with asymmetrical and symmetrical FG cores are illustrated, respectively. Negative velocities in this figure are due to the bouncing back of the projectile after failure of some layers in the sandwich beam. As shown in Fig. 3 the ballistic limit for a beam with an asymmetrical core can be estimated as 135 m/s while based on Fig. 4 this limit for a sandwich beam with a symmetrical core is 129 m/s. Therefore, it can be concluded that the impact strength of a sandwich beam with asymmetrical core is higher than that of one with the symmetrical core.

The velocity variation of projectile ($\Delta V = V_{entry} - V_{exit}$) for different initial velocities that are higher than the ballistic limit is shown in Fig. 5. It can be seen from these figures that for velocities higher than 200 m/s, ΔV in both of sandwich beams with asymmetrical and symmetrical FG cores remains constant. Thus for these velocities, one can calculate the residual velocity of the projectile with knowing its initial velocity [20]. Also it can be concluded that in hyper velocities ($V_i > 200$ m/s) the type of FG core doesn't have any significant effect on the behavior of sandwich beam.

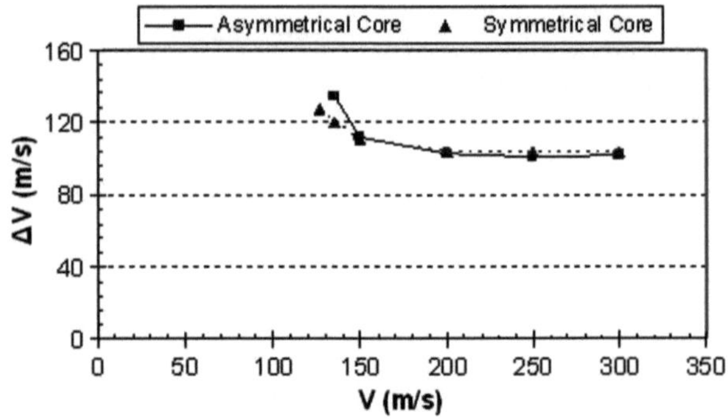

Fig. 5 The velocity variation for asymmetrical and symmetrical FGM core

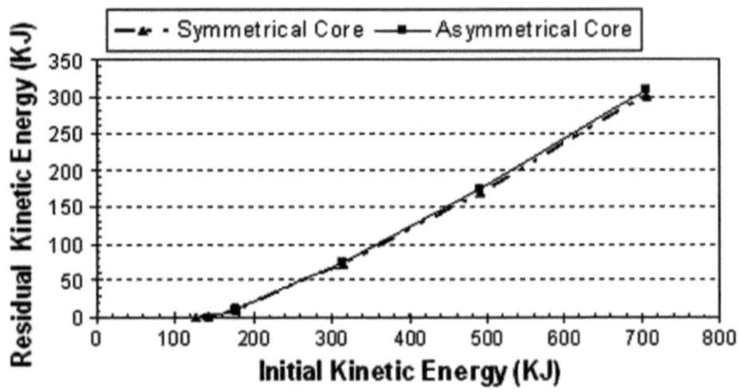

Fig. 6 Residual kinetic energy to initial kinetic energy for asymmetrical and symmetrical FGM core

Figure 6 shows the residual kinetic energy of the projectile versus its initial kinetic energy. As it can be seen from this figure, for two cases of asymmetrical and symmetrical FG cores, variation of these two parameters is linear with the same slope. This fact has been reported for composite structures by Lin and Bhatangar [45]. They investigated ballistic limit and residual velocity of those structures and found a linear relationship for the variation of residual kinetic energy versus initial one. Moreover, the effects of initial kinetic energy of the projectile on its perforation energy passing through the sandwich beam for velocities higher than the ballistic limit is shown in Fig. 7. It is clear that similar linear behavior for both sandwich beams with asymmetrical and symmetrical FG cores was again observed.

4 Parametric Study

4.1 Effects of Failure Strain

In previous section, the failure strain for all FG core layers was assumed to be constant which is not in agreement with reality. Using the same assumptions for material properties of the projectile and sandwich beam, effects of failure strain

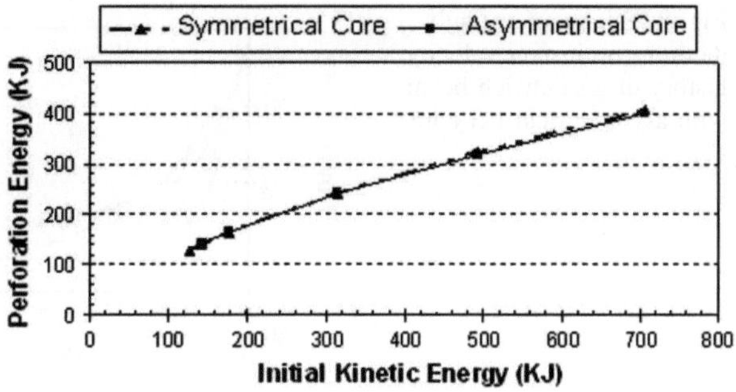

Fig. 7 Perforation energy to initial kinetic energy for asymmetrical and symmetrical FGM core

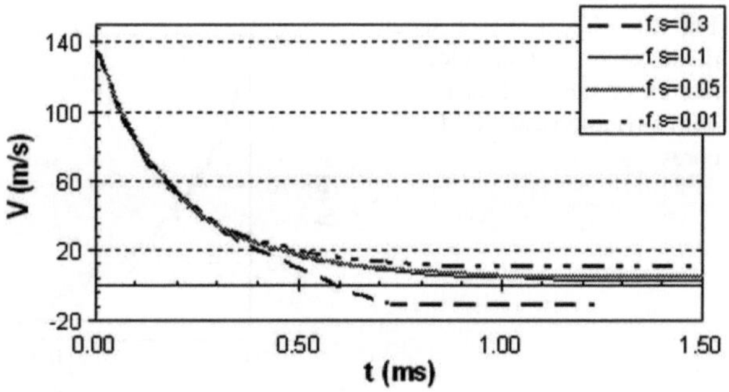

Fig. 8 Velocity history of asymmetrical FGM core for different failure strains

(*f.s*) on instant velocity history of a sandwich beam with an asymmetrical FG core was investigated (see Fig. 8). As it is illustrated in this figure, with increasing the failure strain from 0.01 to 0.3, the impact strength of sandwich beam increases and the projectile cannot penetrate completely in the target.

4.2 Effects of Tangent Modulus

The instant velocity history of a sandwich beam with an asymmetrical FG core for different ratios of Young's modulus to tangent modulus is represented in Fig. 9. Tangent modulus for the face sheets is as same as previous section, i.e. 1 GPa While the ratio of Young's modulus to tangent modulus was 150. In this case a complete penetration was observed and the ballistic limit is estimated at 131 m/s.

4.3 Effects of Thickness of the Face Sheets

The instant velocity history of sandwich beams having face sheets with different thicknesses is shown in Fig. 10. It's clear that with increasing the thickness of face

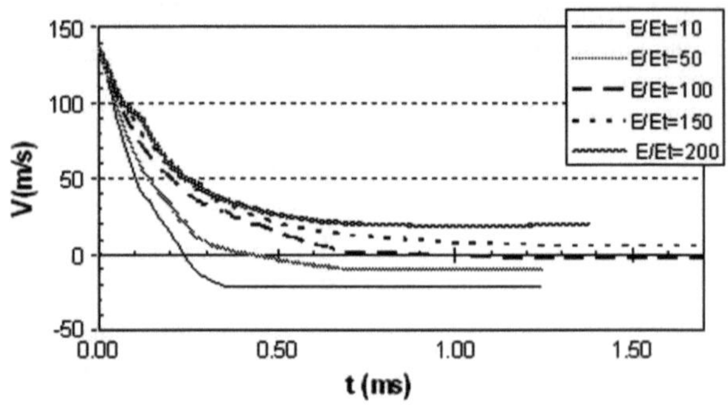

Fig. 9 Effects of young's modulus on instant velocity history of a sandwich beam with asymmetrical FG core

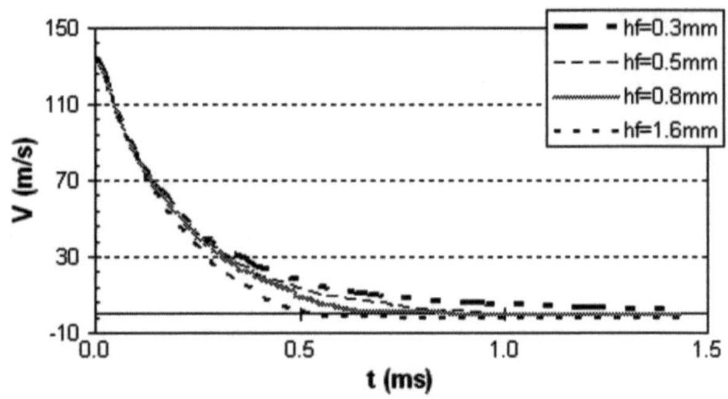

Fig. 10 Effects of face sheet thickness on instant velocity history of sandwich beams with asymmetrical FGM cores

sheets from 0.3 to 1.6 mm, the impact strength of the structure is increasing. But the most important benefit of sandwich structures compared to conventional composites is their specific strength, i.e. strength to weight ratio. Thus with increasing the thickness of face sheets, the total mass of the sandwich beam increases. Therefore, the optimum thickness of the face sheets should be estimated in a design process.

4.4 Effects on Damage Extension

Considering a sandwich beam with an asymmetrical FG core with the ratio of Young's modulus to tangent modulus equals to 150 and failure strain of 0.2, the extension of damage under high velocity impact loading was investigated. The total number of layers for the structure was 43 (2 face sheets + 41 layers at the core). Figure 11 shows the number of failed layers for different initial velocities from 25 m/s up to the ballistic limit of 131 m/s. It is clear from this figure that for velocities higher than the ballistic limit, the number of failed layers will be 43, i.e. full penetration will happen.

In addition, Fig. 12 shows a view of a sandwich beam that failed while impacted with a projectile having an initial velocity of 50 m/s. As it can be seen

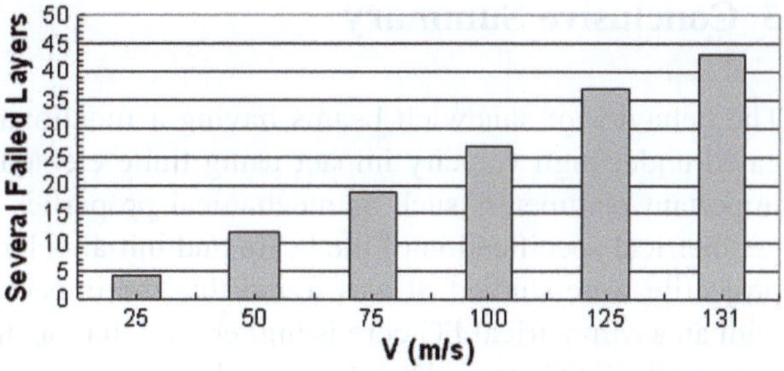

Fig. 11 Number of failed layers at FG core for different initial velocities in a sandwich beam with asymmetrical FG core

Fig. 12 View of failed sandwich beam

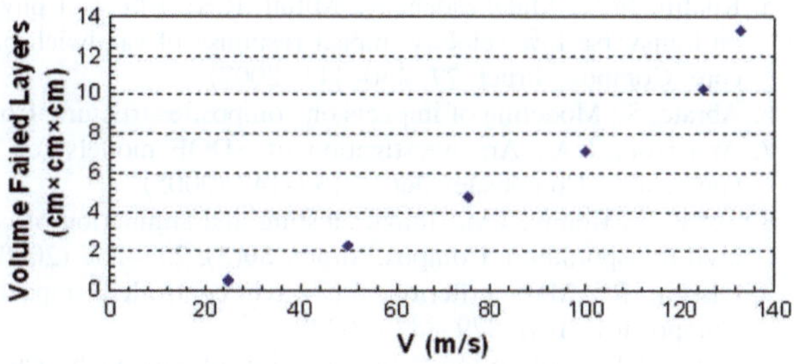

Fig. 13 Failure volume of sandwich beam to different velocities

from this figure, the damage extension can also be quantified based on the failed volume of the sandwich beam. Since the height of the projectile equals to the width of the sandwich beam, so the length of failure in y-direction for every layer of sandwich beam will be equal to its width (0.03 m). To determine the total failed area in x and z directions, failure area of every layer was calculated and summed up. Since only one fourth of the projectile and sandwich beam was modeled, then one forth of total failed area is considered. Figure 13 shows the failed volume of sandwich beam for different initial velocities from 25 m/s up to the ballistic limit of 131 m/s. Similar to the failed number of layers, a nonlinear behavior can be seen in this figure.

5 Conclusive Summary

The behavior of sandwich beams having a functionally graded core was investigated under high velocity impact using finite element simulations. The effects of important parameters such as mechanical properties of the constituent materials, geometrical specification of the beam and initial velocity and kinetic energy of the projectile were studied. It was found that the impact strength of sandwich beams with an asymmetrical FG core is higher than that of similar sandwich beams with a symmetrical FG core. Due to the lake of experimental work in high velocity impact of sandwich beams with FG cores in open literature, this study can be used as an effective tool for designing new experiments.

References

1. Zenkert, D.: The Handbook of Sandwich Construction. Engineering Materials Advisory Services Ltd, London (1997)
2. Lee, L.J., Huang, K.Y., Fann, Y.J.: Dynamic response of composite sandwich plates subjected to low velocity impact. In: Proceedings of the 8th International Conference on Composite Materials, pp. 1–10 (1991)
3. Sun, C.T., Wu, C.L.: Low velocity impact of composite sandwich panels. In: Proceedings of the 32nd Structure Materials Conference, Baltimore, pp. 1123–1129 (1991)
4. Frostig, Y., Thomsen, O.T.: Higher-order free vibration of sandwich panels with a flexible core. Int. J. Solids Struct. **41**, 1697–1724 (2004)
5. Khalili, M.R., Malekzadeh, K., Mittal, R.K.: Effect of physical and geometrical parameters on transverse low-velocity impact response of sandwich panels with a transversely flexible core. Compos. Struct. **77**, 430–443 (2007)
6. Abrate, S.: Modeling of impacts on composite structure. Compos. Struct. **51**, 129–138 (2001)
7. Anderson, T.A.: An investigation of SDOF models for large mass impact on sandwich composites. Composites **36**(2), 135–142 (2005)
8. Torre, L., Kenny, J.M.: Impact testing and simulation of composite sandwich structures for civil transportation. Compos. Struct. **50**(3), 257–267 (2000)
9. Olsson, R.: Mass criterion for wave controlled impact response of composite plates. Composites **31**(8), 879–887 (2000)
10. Ávila, A.F.: Failure mode investigation of sandwich beams with functionally graded core. Compos. Struct. **81**(3), 323–330 (2007)
11. Freeman, B., Schwingler, E., Mahinfalah, M., Kellogg, K.: The effect of low-velocity impact on the fatigue life of sandwich composites. Compos. Struct. **70**(3), 374–381 (2005)
12. Anderson, T., Madenci, E.: Experimental investigation of low-velocity impact characteristics of sandwich composites. Compos. Struct. **50**(3), 239–247 (2000)
13. Schubel, P.M., Luo, J.J., Daniel, I.M.: Low velocity impact behavior of composite sandwich panels. Composites **36**(10), 1389–1396 (2005)
14. Palazotto, A.N., Herup, E.J., Gummadi, L.N.B.: Finite element analysis of low-velocity impact on composite sandwich plates. Compos. Struct. **49**(2), 209–227 (2000)
15. Meo, M., Morris, A.J., Vignjevic, R., Marengo, G.: Numerical simulations of low-velocity impact on an aircraft sandwich panel. Compos. Struct. **62**, 353–360 (2003)
16. Dazhi, J., Dongwei, S.: Local displacement of core in two-layer sandwich composite structures subjected to low velocity impact. Compos. Struct. **71**(1), 53–60 (2005)

17. Choi, I.H.: Contact force history analysis of composite sandwich plates subjected to low-velocity impact. Compos. Struct. **75**(1–4), 582–586 (2006)
18. Gottesman, T., Bass, M., Samuel, A.: Criticality of impact damage in composite sandwich structures. In: Proceedings of 6th Inc. Conference of Composite Materials, pp. 3–27 (1987)
19. Avery, J.L., Sankar, B.V.: Compressive failure of sandwich beams with debonded face sheets. J. Compos. Mater. **34**(11), 1176–1199 (2000)
20. Abrate, S.: Impact on Composite Structures. Cambridge University Press, Cambridge (1998)
21. Charles, J.P., Guerda-Pegeorges, D.: Impact damage tolerance of helicopter sandwich structures. In: Proceedings of 23rd International SAMPE Conference, pp. 51–61 (1991)
22. Palm, T.E.: Impact resistance and residual compression strength of composite sandwich panels. Compos. Eng. **5**(5), 273–286 (1995)
23. Wu, C.L., Sun, C.T.: Low velocity impact damage in composite sandwich beams. Compos. Struct. **34**(1), 21–27 (1996)
24. Shipsha, A., Hallstrom, S., Zenkert, D.: Failure mechanisms and modeling of impact damage in sandwich beams: a 2D approach. Part I: experimental investigation. J. Sandwich Struct. Mater. **5**(1), 7–31 (2003)
25. Venkataraman, S., Sankar, B.V.: Elasticity solution for stresses in a sandwich beam with functionally graded core. AIAA J. **41**(12), 2501–2505 (2003)
26. Nakamura, T., Wang, Z.Q.: Simulations of crack propagation in graded materials. Mech. Mater. **36**(7), 601–622 (2004)
27. Nettles, A.T., Hodge, A.J.: Impact testing of glass/phenolic honeycomb panels with graphite/epoxy faces sheets. In: Proceedings of 35th International SAMPE Symposium CA, pp. 1430–1440 (1990)
28. Sheikh, A.H., Bull, P.H., Kepler, J.A.: Behaviour of multiple composite plates subjected to ballistic impact. Compos. Sci. Technol. **69**(6), 704–710 (2009)
29. Aktay, L., Johnson, A.F., Holzapfel, M.: Prediction of impact damage on sandwich composite panels. Comput. Mater. Sci. **32**, 252–260 (2005)
30. Bohong, G., Jingyi, X.: Finite element calculation of 4-step 3-dimensional braided composite under ballistic perforation. Compos. B **35**(4), 291–297 (2004)
31. Kärger, L., Baaran, J., Tebmer, J.: Rapid simulation of impacts on composite sandwich panels inducing barely visible damage. Compos. Struct. **79**(4), 527–534 (2007)
32. Bashurov, V.V., Bebenin, G.V., Belov, G.V., Bukharev, Y.N.: Experimental modelling and numerical simulation of high- and hypervelocity space debris impact to spacecraft shield protection. Int. J. Impact Eng. **20**(1–5), 69–78 (1977)
33. Ambur, D.R., Jaunky, N., Lawson, R.E., Knight, N.F.: Numerical simulations for high-energy impact of thin plates. Int. J. Impact Eng. **25**(7), 683–702 (2001)
34. Anderson, T.A.: A 3-D elasticity solution for a sandwich composite with functionally graded core subjected to transverse loading by a rigid sphere. Compos. Struct. **60**(3), 265–274 (2003)
35. Apetre, N.A., Sankar, B.V., Ambur, D.R.: Low-velocity impact response of sandwich beams with functionally graded core. Int. J. Solids Struct. **43**(9), 2479–2496 (2006)
36. Kashtalyan, M., Menshykova, M.: Three-dimensional elasticity solution for sandwich panels with a functionally graded core. Compos. Struct. **79**, 587–892 (2007)
37. Kirugulige, M.S., Kitey, R., Tippur, H.V.: Dynamic fracture behavior of model sandwich structures with functionally graded core: a feasibility study. Compos. Sci. Technol. **65**(7–8), 1052–1068 (2005)
38. Etemadi, E., Afaghi, A., Takaffoli, M.: 3D finite element simulation of sandwich panels with a functionally graded core subjected to low velocity impact. J. Compos. struct. **89**, 28–34 (2009)
39. Das, M., Barut, A., Madenci, E., Ambur, D.R.: A triangular plate element for thermo-elastic analysis of sandwich panels with a functionally graded core. Int. J. Numer. Meth. Eng. **68**, 940–966 (2006)
40. Zhu, H., Sankar, B.V.: Analysis of sandwich TPS panel with functionally graded foam core by Galerkin method. Compos. Struct. **77**, 280–287 (2007)

41. Lebe, P., Dong, M., Schmauder, S.: Self-consistent matricity model to simulate the mechanical behavior of interpenetrating microstructures. Comput. Mater. Sci. **15**(4), 455–465 (1999)
42. Tohgo, K., Masunari, A., Yoshida, M.: Two-phase composite model taking into account the matricity of microstructure and its application to functionally graded materials. Compos. A **37**(10), 1688–1695 (2006)
43. Chan, S., Fawaz, Z., Behdinan, K., Amid, R.: Ballistic limit prediction using a numerical model with progressive damage capability. Compos. Struct. **77**(4), 466–474 (2007)
44. Naik, N.K., Shrirao, P.: Composite structures under ballistic impact. Compos. Struct. **66**(1–4), 579–590 (2004)
45. Lin, L.C., Bhatangar, A.: Ballistic energy absorption of camposites. Int. SASMPE Tech. Conf. **23**, 669–683 (1991)

Multiscale Modeling of the Effect of Very Large Strain on the Microstructure Evolution and Ductility of Microalloyed Steels

Krzysztof Muszka and Janusz Majta

Abstract This study addresses some aspects regarding mechanical behavior of bcc structures characterized by high grain refinement level developed using large plastic deformation. The fundamental mechanisms governing the behavior of the microalloyed steels at wide range of deformation conditions at nano, micro, meso-meter scale and at the continuum are discussed. Grain refinement levels, where the change in the ability of grains to strain hardening is observed, are identified with respect to the resulting changes in the meso and macro levels effects of precipitation and solid solution strengthening mechanisms. Differences in the description of strengthening mechanisms and ductility represented by maximum uniform elongation of ultrafine-grained (UFG) and nanocrystalline materials are also defined. Existing flow stress models for UFG materials are presented and their physical bases are discussed with respect to their application in the computer modeling process of mechanical behavior of bcc structures strengthened by alloying elements.

Keywords Microalloyed steel · Severe plastic deformation · Ultrafine-grained structure · Multiscale modeling

1 Introduction

Presented study includes experimental findings and the present theories as well as physical and computational modeling required to understand the correlations between microstructure evolution and mechanical behavior of metals subjected to large plastic deformations. Presently, the real challenge for economically and reliably predicting and designing materials for the specific use at these very

K. Muszka (✉) · J. Majta
AGH University of Science and Technology, Mickiewicza 30, 30-059 Krakow, Poland
e-mail: muszka@agh.edu.pl

J. Majta
e-mail: majta@metal.agh.edu.pl

N. Bonora and E. Brown (eds.), *Numerical Modeling of Materials Under Extreme Conditions*, Advanced Structured Materials 35, DOI: 10.1007/978-3-642-54258-9_6,

complex deformation conditions is to properly build multiscale models. Microalloyed steels due to a very good strength-to-ductility balance represent one of the most competitive structural material with the wide field of applications. Thanks to their microstructural complexity, where grain size control, substructure, precipitation and solid solution strengthening mechanisms play a significant role, they also represent very interesting research material. Nowadays, when strong grain refinement is recognized as a tool for further improvement of the mechanical properties its combination with the beneficial role of microalloying elements seems to be promising route to achieve good quality product characterized by both good ductility and strength. Recently developed severe plastic deformation (SPD) techniques have made it possible to produce submicron/nanostructures also in microaloyed steels. However, these new fields of modern materials manufacturing need to be controlled by highly developed multiscale rheorogical and microstructural models. These models must be capable of predicting deformation and strengthening mechanisms at critical locations in the process course and over a wide range of materials. Hence, the real challenge for the state-of-the-art of modern materials behavior modeling is the process of microalloyed steel forming, especially under SPD conditions. There is a growing market for ultrafine-grained (UFG) microalloyed steels, and the factors that encourage this expansion are numerous and complex. New experimental and in situ diagnostic techniques coupled with advanced microstructure analysis (TEM, EBSD) have revealed a mechanical phenomena with a new field of potential applications. This necessitates a detailed knowledge of the constitutive response of UFG materials. In the present study the ultrafine-grained microstructures are produced by multiaxis compression tests through a process called continuous recrystallization or in situ recrystallization [2, 24]. A number of authors have studied and modeled these processes in metals and alloys containing second phase particles processed by SPD [1, 13, 14, 31, 36]. However, most of the works were focused on nonferrous metals and alloys. The ductility of the ultrafine-grained microstructures decreases as an effect of the decrease in strain hardening i.e. the tensile strength becomes very close to the yield strength. It is proposed by Considére [4] that the plastic instability condition in tensile tests, the moment when the material is in unstable equilibrium, can be expressed by the formula $\sigma > \mathrm{d}\sigma/\mathrm{d}\varepsilon$, where σ is the flow stress and ε is the true strain. Hence, we can expect that using second phase particles it is possible to improve the strength-ductility balance in ultrafine-grained microalloyed steels. Also it is observed in bcc metals that the smaller grain size materials are much stronger at low strain rates, but show less relative strengthening at high rates [15, 34]. Again, presence of fine dispersed second phase particles could be beneficial from point of view of strain hardening at higher strain rates, leading to improved ductility under dynamic loading conditions. In the case of ultrafine-grained microalloyed steels there is no grain growth evidence at temperatures up to about 650 °C. Similarly, in the present study, after annealing at 500 °C for 1,200 s of the deformation induced ultrafine structure the grain size distribution was still nanoscale and the only effect observed was the aid of recovery-continuous recrystallization. Also here we can believe that the inhibition of grain growth is due to

Table 1 Basic chemical composition of the investigated steels (in wt.%)

Steel	C	Mn	Si	Al	Ti	Nb	B
A	0.07	1.36	0.27	0.02	0.031	0.067	0.0002
B	0.0022	0.11	0.009	0.037	0.073	–	–

salute drag on the grain boundaries and due to pinning of the boundaries by the nanoscale intermetallic precipitates.

Computer modeling is an effective tool to analyze the microstructure evolution in materials subjected to SPD at different scale levels. Application of the Finite Element Method (FEM) for calculations and modeling of the mechanical response of UFG materials needs properly built rheological models [27, 28]. In the case of severely deformed microstructures that are mostly characterized by high inhomogeneity of microstructure and mechanical properties, existing flow stress models need to be modified and justified to the new conditions. The proper analysis based on the mechanism of generation the dislocation structures and taking into account effects of precipitation and solid solution strengthening, implemented in FEM, enable to capture the moment of plastic instability of UFG microalloyed steels. The fact that ductility, strengthening and deformation mechanisms are strictly connected to each other suggests that the properly built model of their relationships is the key factor in the improving process of UFG materials.

Presented results introduce a spectrum of experimental and theoretical investigations that systematically quantify the influence of variables such as strain rate, temperature, microstructure, and alloying content on the constitutive behavior, dislocation substructure formation and evolution. Physical and computer modeling were employed and explored, to analyze and predict the behavior of the materials aided by dislocation theory in deformation mechanisms and a new model of flow stress in application to ultrafine-grained materials. The chapter encompasses a range in scale of the problems at hand starting from macro, meso, micro, nano and atomistic scale.

2 Experimental

This work is focused on studies promoting understanding of the multiphase and ultrafine-grained material properties related to the behavior observed in typical microstructures. The aim of the experimental part of the present research was to obtain a wide range of microstructures with different levels of grain refinement in order to study the effect of grain refinement and their mechanical behavior as well as ductility. The basic chemical compositions (wt.%) of the investigated steels are presented in Table 1. Both steels were supplied in the "as hot rolled" condition.

The details concerning the experimental work can be found in [24, 29]. The main processing parameters employed in the present study, are presented in Table 2.

Table 2 Deformation schedules

Route	T_D (°C)	ε	T_A (°C)	t_A (s)	CR_1 (°C/s)
1	20	2	500	1,200	4
2a	20	5	–	–	–
2b	20	7	500	1,200	4
3	20	10	500	1,200	4
4	20	15	500	1,200	4
5	20	20	500	1,200	4

The specimens made from microalloyed steel (A) were deformed in multiaxial compression (MaxStrain) with the wide range of total strains from 2 to 20 (strain of 0.5 per each pass). In most cases, after deformation, specimens were reheated to the annealing temperature (500 °C) and annealed for 1,200 s. Additionally, in deformation route 2a, specimen was not annealed in order to provide more data for the modeling process of substructure strengthening. As a reference material, interstitial free steel (B) was also used—however this material was only subjected to the deformation route of 5.

Mechanical behavior of the developed microstructures was studied using tensile tests. Microstructure evolution was studied utilizing of Transmission Electron Microscopy (TEM), whilst the volume fraction of LABs and HABs was estimated using Electron Back Scattered Diffraction (EBSD) analysis performed for the specimens after MaxStrain deformation.

3 Results

In ultrafine-grained steels produced by SPD techniques, the formation of an ultrafine microstructure can be attributed to very large accumulated strain and deformation temperature as well as to the heat treatment that was applied to the final products. In the case of microalloying steels fine precipitates as well as alloying elements in solution play also very important role. To examine the microstructure evolution EBSD analysis were performed in the present study (Fig. 1). Microstructure observation revealed that in general in schedule 1 (MaxStrain), a much higher volume fraction of LABs can be noted for steel A. This suggests a much stronger tendency for this steel to generate the dislocation structures in the form of cells and subgrains. The dispersed precipitates of (Nb, Ti) (C, N) in this steel effectively pin dislocations so the effects of deformation in severe multiaxis compression are accumulated. Continuously increasing in microstructure refinement can be observed in Fig. 1. Comparing Fig. 1a ($\varepsilon = 2$) and 1f ($\varepsilon = 20$), it can be seen that when the total strain increases the steel A "produces" a higher volume fraction of LABs, that can be transformed into HABs under certain favorable conditions (e.g. after applying subsequent annealing). From Fig. 1 it is seen that microstructure after schedules 3 and 4 are still quite

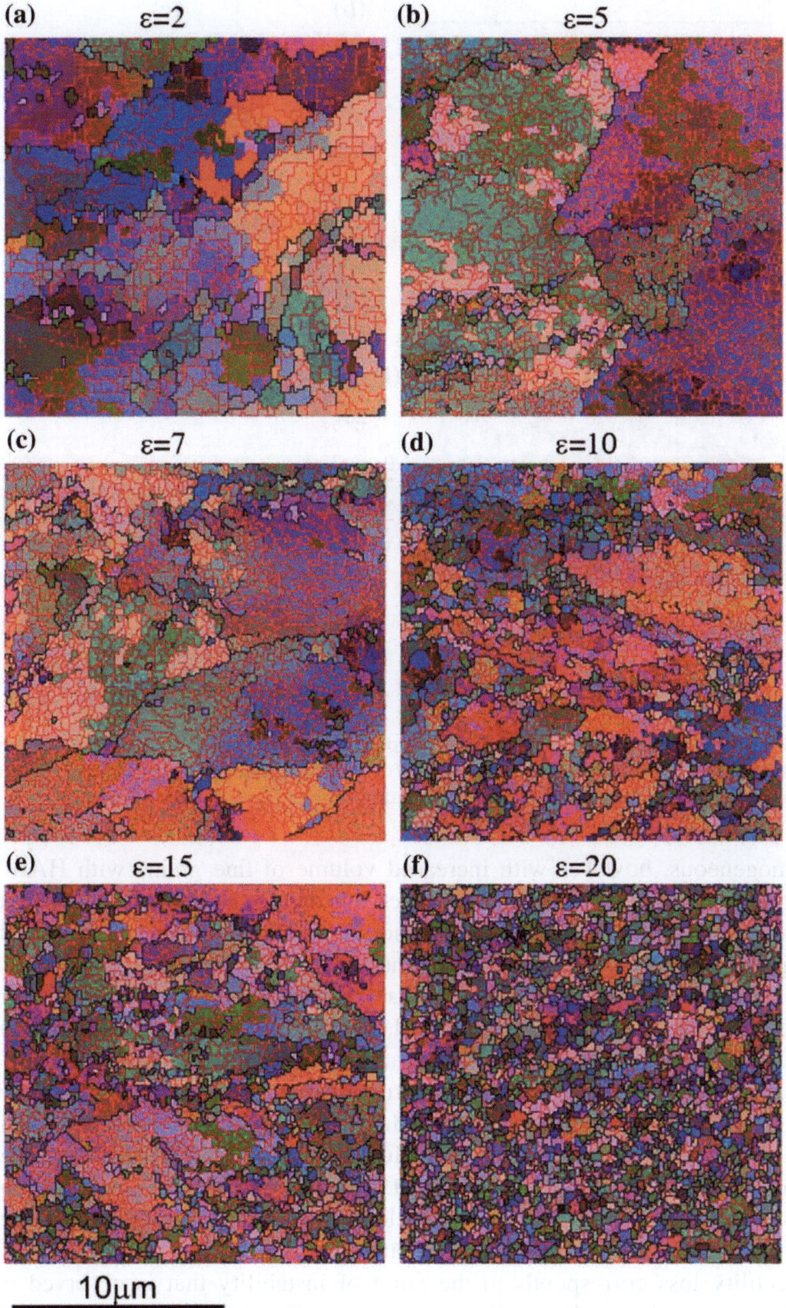

Fig. 1 Electron backscatter diffraction results showing development of grain boundaries with increasing strain

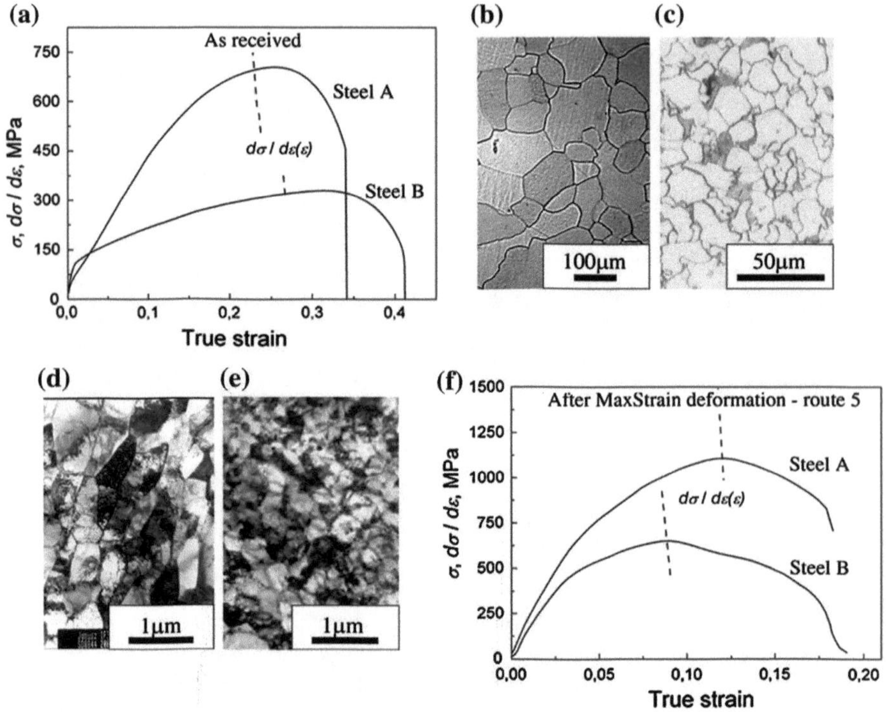

Fig. 2 Determination of the uniform elongation in: coarse-grained (**a**), and ultrafine-grained materials (after *SPD* MaxStrain deformation) (**f**). TEM of initial structures of steel B (**b**) and A (**c**), and structures deformed with the route 5 from Table 2: steel B (**d**) and A (**e**)

inhomogeneous, however with increased volume of fine grains with HAB. This inhomogeneity is almost diminished after schedule 5. In this sample well-recovered refined grains with clean grain interior dominate the microstructure, as shown in Fig. 1f ($\varepsilon = 20$). Besides of the volume fraction of the precipitates, their size is also important in terms of interactions with the dislocations. Also from the authors' recent study [29], it is believed, that the refined dislocation structure of microalloyed steel results from the interaction with fine precipitates, which forces dislocations to form the fine local cell structures, in contrast to the situation where the uniform dislocation lines propagate due to the lower number of precipitates. It is found in the present study that the volume fraction of precipitated carbides has an effect on the proportion of different types of cells. The higher the volume fraction, the lower the relative ratio of closed cells to open cells—the dislocation substructure is similar to those obtained after low temperature deformation [5, 19].

Ductility loss corresponds to the point of instability that is observed on the tensile curve as the end of uniform elongation. One of the main reasons of low ductility in ultrafine-grained materials can be attributed to the dynamic recovery, that causes the lower strain hardening rate in these materials [12, 35]. Figure 2 shows examples of microstructure analysis and the tensile curves recorded for ultrafine-grained materials obtained in microalloyed (A) and interstitial free (B) steels using MaxStrain deformation (route 5 from Table 2). These curves are compared with those measured for typical as-hot-rolled material. It is clearly confirmed that the significant grain refinement causes adequate increase in strength. Also the decrease in ductility can be easily observed. The Considére criterion is applied here to find the moment of plastic instability in the tension tests. The uniform elongation significantly decreased with grain refinement in both deformed materials and reached value below 0.10 after deformation in case of steel B, and below 0.12 in case of UFG steel A [23]. It should be noticed that after SPD steel A represents, in contrast to the initial materials, better ductility represented here by the uniform elongation.

It can be seen that for the same SPD conditions (Table 2—schedule 5) steel A, where content of the alloying elements was higher compared to steel B, shows much stronger grain refinement (Fig. 2d, e). In this case, also the degree of HABs generation (misorientation angle >15°) is slightly more advanced.

There are several ideas that can be used to improve ductility in ultrafine-grained materials [3, 12, 20]. First of all, due to Considére criterion, by increasing the ability of ultrafne-grained material to strain hardening. Very effective could be also creating a bimodal structure (inhomogenous microstructure consisting of both ultrafine and typical grains) that would assure both good ductility and balance between the strength and ductility. Another route involves introduction of hard second phase (martensite or cementite) to promote dislocation activity through the dual phase effect. Finally, as it is observed in microalloyed steels discussed in the present study, increased uniform elongation can be achieved by introducing disperse Nb carbonitrides and by solid solution strengthening. It is clearly shown by the experimental results given above that increased ductility in UFG microalloyed steels is an effect of dislocation-second phase particles interactions during severe plastic deformation, evidenced by the appearance of increased dislocation density and finally large amount of fine grains with high angle grain boundaries as a result of continuous recrystallization.

4 Modeling

The most important elements in the modeling process of UFG ferrite microstructure evolution and ductility under conditions of very large strains are the chemical composition and amount of accumulated deformation. From point of view of plastic flow, for slip to occur, the dislocation must either move around the particles or through the particles. The dislocations generated during deformation may avoid the particles or obstacles by leaving the slip plane in the vicinity of each particle, or it may avoid the particles. In either case, the internal energy increases and the dislocation substructure develops, however in different intensity. The strengthening mechanisms operate in UFG microalloyed steels, produced by SPD processes are of great interest. This chapter contains a brief overview of strengthening components of the flow stress model from the classical (Hall-Petch) and dislocation substructure evolution (Hansen) perspective. It is not the intension of this chapter to discuss all possible aspects of rheological models of UFG microalloyed steels, but rather to concentrate on the mechanisms by which dislocation substructure and finally grain refinement occur during deformation to extreme plastic strains.

Nucleation theory, diffusion-controlled growth and the solubility product of niobium, vanadium, titanium, carbon, and nitrogen in the austenite of microalloyed steels have been very often used to derive equations for the start of precipitation. The kinetics of the precipitation process is also clearly influenced by the extent of deformation and the precipitation process occurs at progressively shorter times, as the deformation is increased. However, in real metal forming conditions the interpass times are too short for a full precipitation process, so the amount of strain-induced precipitates is considerably less than expected from the chemical composition. For example, Sellars in one of his fundamental works [33] stated that each of the stages of reheating, roughing, and finishing should be considered in turn to examine the particle interactions. It was also shown that the minimum particle size in equilibrium at a given temperature must be 50 % larger for particles in the matrix than for particles on grain boundaries. Thus, particles in this size range, which precipitated on grain boundaries, will become unstable and redissolve if the boundary migrates away from them. However, Sellars has also concluded that detailed study of the particles in microalloyed steels is difficult because fine particles of 2–3 nm in diameter are critical, but cannot be observed on extraction replicas, and are difficult to observe in thin foils. This is because of the high dislocation density produced by the martensite transformation during the rapid quench usually used to preserve the particle and grain structure existing at high temperatures. It means that more basic research is still required to establish physical-based model of the complex interactions during hot working but also cold working processes of niobium-microalloyed steels. Any strain-induced precipitation in the austenite will naturally reduce the amount of niobium that is available for subsequent interphase precipitation during austenite to ferrite transformation or during cooling in the ferrite. Hence, the microstructures of microalloyed steels formed in the austenite as well as during and following the austenite-ferrite transformation should be the starting point in the analysis and modeling of SPD effects.

4.1 Development of the Flow Stress Model for UFG Materials

For the several decades, the basic relationship between the grain size and the yield stress has been well described by equation introduced by Hall and Petch [7, 32]:

$$\sigma_p = \sigma_0 + k_y d^{-1/2} \tag{1}$$

where: σ_p—yield stress, σ_0—internal friction, k_y—coefficient, d—grain size. However recently, when strong grain refinement (grain size below hundreds of nanometers) can be obtained, some deviations from the prediction given by Hall-Petch equation are observed. According to Conrad [3] who examined the influence of the rate controlling mechanisms of different metals over the grain size range from tens of nanometers to several millimeters, the existence of the three regimes, where a change in the Hall-Petch slope (due to change in material behavior) exists, can be postulated. The critical grain size indicating transition from one regime to another depends on both processing parameters (temperature, strain, strain rate) and material features (initial microstructure and texture). The grain size strengthening for coarse-grained material (i.e. the Hall–Petch relation—see Eq. 1) can be well described by the dislocation pile-up model [literature]. However, the pile-up model is no more successful in explaining the flow stress of UFG materials where the high dislocation density is introduced. Therefore, new models describing strengthening mechanisms are proposed. Recently, flow stress of UFG pure titanium was proposed to be described by the dislocation bow-out model [18]. Regarding the physical model based on dislocation bow-out behavior, the yielding is closely related to the critical shear stress required to make a semicircle configuration of the Frank–Read source.

Theoretically, the critical stress for such condition is approximated by the expression in Eq. (2) [10]:

$$\tau_{cr} = \frac{Gb}{2\pi L(1-\nu)}\left[\left(1-\frac{3}{2}\nu\right)\ln\left(\frac{L}{b}\right) - 1 + \frac{\nu}{2}\right] \tag{2}$$

where G—shear modulus, b—Burgers vector, ν—Poisson's ratio and L—the average dislocation length. For grain sizes larger than 100 nm, it can be assumed that L is equal to $\rho^{-1/2}$.

Also Nes et al. [30] indicated, that in the case of microstructures with the submicron grain size, due to the fact that the spacing of mobile dislocations is comparable to the grain size and when a stress is applied, these dislocations will be forced to bow out the boundary—stable pile-up configurations do not form, so this mechanism cannot be used to explain the Hall-Petch relationship. The authors proposed that the stress required for dislocation migration in a substructure consisting of a mixture of LABs and HABs with a Frank network of dislocations inside the subgrains/grains is represented by:

$$\tau = \tau_t + \hat{\tau}_p + \alpha_1\, Gb\sqrt{\rho_i} + \alpha_2\, Gb\left(\frac{1}{\delta} + \frac{1}{D}\right) \tag{3}$$

where G is a shear modulus, b is Burgers vector, $\hat{\tau}_p$ is the flow stress contribution caused by non-deformable particles (Orowan by-pass stress $\hat{\tau}_p = Gb/\lambda$, where λ is the particles spacing), ρ_i is the dislocation density in the subgrain interior and δ is the separation of the LABs. D is either the grain size in un-deformed polycrystalline metals or the separation of HABs in heavily deformed metals.

However, other work suggests that the pile-up model is still valid. Hansen [9] explained the strengthening caused by LABs based on a dislocation pile-up, where the boundary resistance was taken as the stress to push the edge dislocations through a simple tilt wall and proposed a Hall-Petch slope to be approximately equal:

$$k = G\sqrt{b\theta} \tag{4}$$

where θ is the misorientation angle. As an alternative explanation Hansen related the flow stress of a cell-forming metal to the total dislocation density. He assumed that the dislocation density for a mixed tilt/twist boundary only results from geometrically necessary dislocation boundaries and proposed the following flow stress relation:

$$\tau = \tau_0 + M\alpha Gb\sqrt{3b\theta}D_B^{-1/2} \tag{5}$$

where τ_0 is a friction stress, M is Taylor factor, D_B is the boundary spacing measured along random lines, G, b—meaning as in Eq. (3). In the case of steel, for a value of M = 3 and $\alpha = 0.24$, the Hall-Petch slope can approximately equal:

$$k = 1.25\, G\sqrt{b\theta} \tag{6}$$

This value is higher compared to that obtained from the pile-up model. As mentioned above, Eq. (6) shows the effect of misorientation angle on the Hall-Petch slope for dislocation boundaries. It can be seen that the slope in the Hall-Petch relationship—Eqs. (4) and (6)—for dislocation boundaries increases proportionally to the square root of the misorientation angle. Therefore, at small angles the Hall-Petch slope for cells or subgrains will be lower than in the case of HABs. On that basis, it can be assumed that there exists a critical grain boundary misorientation angle at which the strengthening of the LABs and HABs are equal. The value of this critical angle changes for different materials and for example for IF steel is equal to 0.52° [9]. Therefore, it can be summarized that the mechanical response of ultrafine-grained steels can be treated as a combined effect of different dislocation arrangements. The influence of solid solution and precipitation is accounted for in this analysis by the retardation effect on the dislocation movement. As a consequence, the formation process of LABs is effectively altered. An attempt of substructure strengthening description has been also proposed by the Authors of present work elsewhere [25].

The contribution from LABs and HABs to the strengthening mechanism can be taken into consideration not only by the determination of their separation but also by specifying their density. In the case of severely deformed structures, that are characterized both by dislocation substructures and HABs, their contribution to the strength can be expressed in an additive form [8]:

$$\tau = \tau_0 + M\alpha Gb\sqrt{1.5bS_V\theta_{LAB}(1-f)} + k\sqrt{\frac{S_V}{2}f} \tag{7}$$

where S_V is the area of boundary per unit volume, f is the density of HABs and θ_{LAB} is the average misorientation angle of LABs.

A good prediction has been found using this equation for the analysis of the experimental flow stresses of cold rolled Al, Ni and IF steel. In this equation, S_V can be replaced with $2/D_B$, where D_B is the average boundary spacing. On the basis of the above analysis it can be concluded that the main differences affecting the description of constitutive laws of the strengthening mechanisms of ultrafine-grained and nanostructured structures are a result of the presence of significantly increased volumes of both LABs and HABs that act as obstacles for dislocations. Hence, it can be summarized that the ability to predict the mechanical properties of such microstructures depends on properly established representations of these phenomena in the constitutive equation of the flow stress. The problem is especially complicated in the case of microalloyed steels—the deformation process needs more energy, however when the accumulated deformation energy is high enough, the microalloying elements can be beneficial for the ultrafine-grained microstructure development. The dislocations responsible for creating the cells and subgrains and after annealing the ultrafine-grained microstructures interact with other microstructural features, such as precipitates, impurity atoms and grain boundaries. Unfortunately, in the case of microalloyed steels no clear methodology exists for transferring this knowledge into a proper macroscopic model.

In order to use computer simulation to model mechanical behavior and assess the uniform elongation of UFG materials, it is necessary to use a proper rheological model, that takes into consideration contribution from substructure strengthening. Based on above-mentioned study of the substructure strengthening a modification of the Khan-Huan-Linag flow stress model was proposed, so it relates the strain hardening with strain, strain rate and dislocation cell size [16, 17, 26, 29]:

$$\sigma_p = \sigma_{gb}\left[1 + B^*\left(1 - \frac{\ln\dot{\varepsilon}}{\ln D_{p0}}\right)^{n_1}(\varepsilon^p)^{B^* n_0}\right]\left(\frac{\dot{\varepsilon}}{\dot{\varepsilon}_r}\right)^C\left(\frac{T_m - T}{T_m - T_r}\right)^m \tag{8}$$

where σ_p—flow stress (MPa); σ_{gb}—grain boundary strenghtening (MPa); ε^p—equivalent plastic strain; $D_{p0} = 10^6$ s^{-1}; ε_r—equivalent plastic strain rate (s^{-1}); ε—equivalent plastic strain rate (s^{-1}); T, T_m, T_r,—deformation, melting and reference temperature (K); B, n_0, n_1, C, m—model parameters.

In the present study, further modification of the KHL model was proposed by replacing parameter B with $B^* = B/\sigma_{gb}$. At the same time, parameter B* was introduced to the power function of strain. This modification allowed better

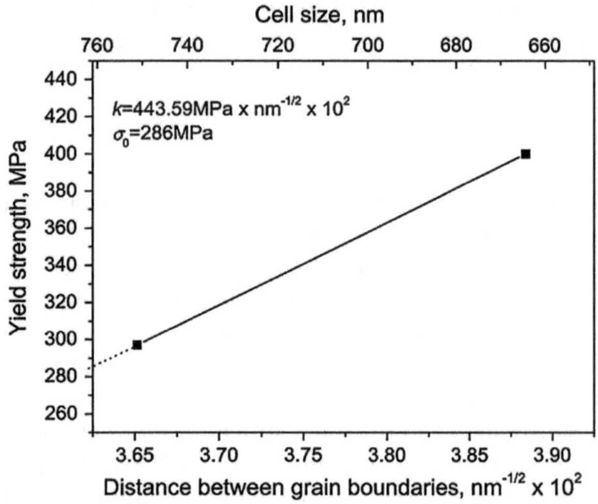

Fig. 3 Way of finding k and σ_0 values for investigated steel A

Table 3 Coefficients of modified KHL model for steel A identified using inverse method

M	α	G (MPa)	b (nm)	σ_0 (MPa)	k (MPa $nm^{-0.5}$)	B	n_1	n_0	C
2.75	0.24	81,700	0.248	286	44,359	5,300	0.1	0.079	0.27

convergence with the experimental results. Basing on the above-mentioned discussion and assumption, that severe plastic deformation leads to dislocation structure development, it was assumed that grain boundary strengthening can be treated as a sum of low- and high angle boundaries—in the same way as in Eq. (7):

$$\sigma_{gb} = \left(\sigma_0 + M\alpha Gb\sqrt{1.5bS_V\theta_{LAB}(1 - f_{HABs})} + k\sqrt{\frac{S_V}{2}f_{HABs}}\right) \tag{9}$$

The values of parameters d_b, θ_{LABs} and f_{HABs} were measured using EBSD analysis, whereas the values of M, α, G, b were taken from the literature.

The values of σ_0 and k were found using a way presented on Fig. 3. The rest of the parameters were identified using inverse method based on the results from the tensile tests (load vs displacement). The values of the optimal model parameters identified are summarized in Table 3.

4.2 Mulitscale Modeling of UFG Microstructures

Often in finite element analysis, the finite element mesh may be too coarse to produce satisfactory results in a region of interest [6, 21, 22]. To obtain more accurate results in such a region, submodeling is an effective finite element

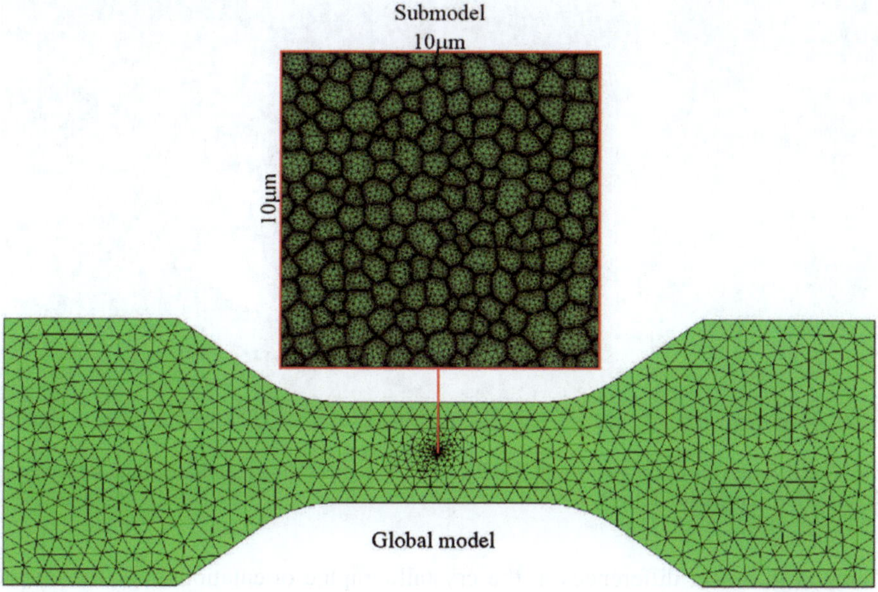

Fig. 4 General idea of multiscale modeling using global and local (submodel) FEM models

technique that can be successfully applied in the process of mulitscale modeling of the mechanical behavior of severely deformed structures. Approach used in the present study is one of the concurrent multi scale computing methods where one strives to solve the problem simultaneously at several scales by an a priori decomposition. Two-scale methods, whereby the decomposition is made into coarse scale and fine scale, have been considered so far. In the general use of submodeling technique, the first step is to solve a problem with one mesh describing the global domain. Then another and generally finer mesh is used to reanalyze a certain sub part of the global domain which is of particular interest. Thus, either the displacement field or the stress field of a global domain is interpolated to the cutting boundary of a sub model having a fine mesh along the edges. In this study, global model with coarse mesh was prepared to model 2D tensile test on flat specimens. Abaqus Standard software was used, with 3-nodes linear plane strain elements type. The submodeling technique was applied in the analysis as shown on Fig. 4. Then the submodel was built, which involved the interested local features, i.e. microstructures developed during severe plastic deformation. Microstructures were designed using Digital Microstructure Representation (DMR) approach [21, 22]. The digital microstructures representing extremely different grain refinement levels were chosen (deformation routes 3a and 5)—Fig. 5—and attached to the central location in the global model—as illustrated in Fig. 4. Flow stress model used was modified KHL model that was implemented using UMAT subroutine. The parameters of the identified material models applied in the submodels were additionally diversified using the Gauss distribution

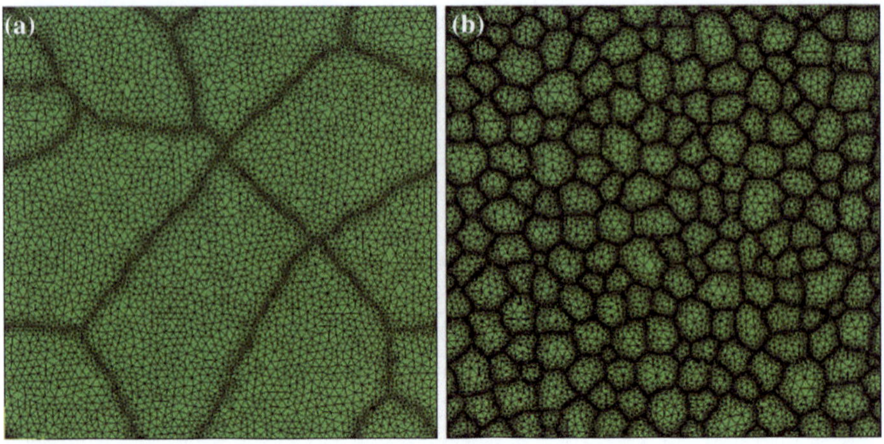

Fig. 5 FEM meshes developed using DMR technique representing microstructures subjected to deformation schedules 2a (**a**) and 5 (**b**)

function to reflect differences in the crystallographic orientations. For the tensile test simulations the same models as for the optimization procedure (see previous paragraph) were used. For the visualization of the Considére criterion, separate state variable was assigned to each mesh element in global model.

At the beginning of simulation that variable was equal to zero. Through the whole simulation process that variable was controlling whether Considére criterion is fulfilled and when the value of strain hardening became lower than the flow stress, state variable was changed to 1.

Another state variable was used to calculate the value of dislocation density of ultrafine-grained structures using Holt model [11]. According to Holt, unstable dislocation arrangement, resulting from periodical fluctuations of the dislocation density, is responsible for the subgrain formation process. Similarly to the spinodal decomposition theory, it can be proved that this instability can increase and lead the dislocations to assemble in the low energy cell arrangements (when screw dislocations acting in parallel are present). Key assumption of this model is the existence of the critical distance, at which dislocations' interactions can be neglected. Then, dislocation cell size can be determined using the wavelength of the fastest growing fluctuation, which leads to the following equation:

$$D_B = k_c \rho^{-1/2} \tag{10}$$

where: D_B—size (diameter) of the dislocation cell, k_c—constant, ρ—dislocation density. Relation between the flow stress and cell size can be derived by substituting Eq. (10) into the Taylor relationship [37]:

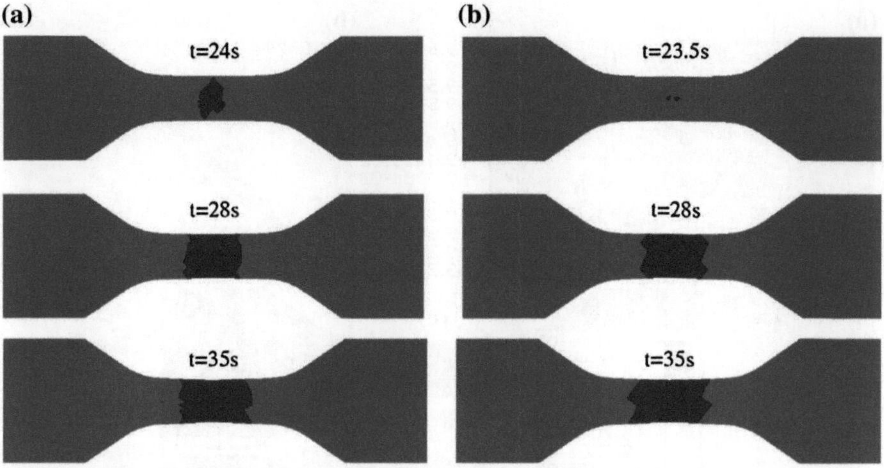

Fig. 6 Onset of necking and its development in subsequent deformation steps. Deformation routes 2a (**a**) and 5 (**b**). *t*—time from the beginning of deformation. Steel A

$$D_B = \alpha k_c b (G/\sigma)^1 \tag{11}$$

Equation (11) can be used to find the cell size that can be utilized then to calculate the dislocation density using Eq. (10). Since in the present study, the dislocation cell size, volume fraction of low angle boundaries, and average misorientation angle of LABs were measured based on the results of EBSD analysis, presented model was used as a postprocessor to find the value of dislocation density.

4.3 Modeling Results

Figure 6 shows the onset of necking in tensile test simulations for A steel after MaxStrain deformation with the routes 2a and 5. Blue area represents mesh elements where the values of the work hardening rate that are below actual values of stress. These elements are related to the places in the material where the onset of necking has just occurred. As it can be seen in the subsequent time steps, these regions are growing as the necking develops.

Equivalent plastic strain distributions for both global and submodels models recorded at the time of onset of necking are presented on Fig. 7. In all the cases it can be seen that results given by submodel show inhomogeneity of the strain at the microstructure level what has not been caught by the global model. As it can be seen in Fig. 7b and f, strain localization starts to develop in the microstructure along the grain boundaries and has a nonuniform character. Areas with narrow

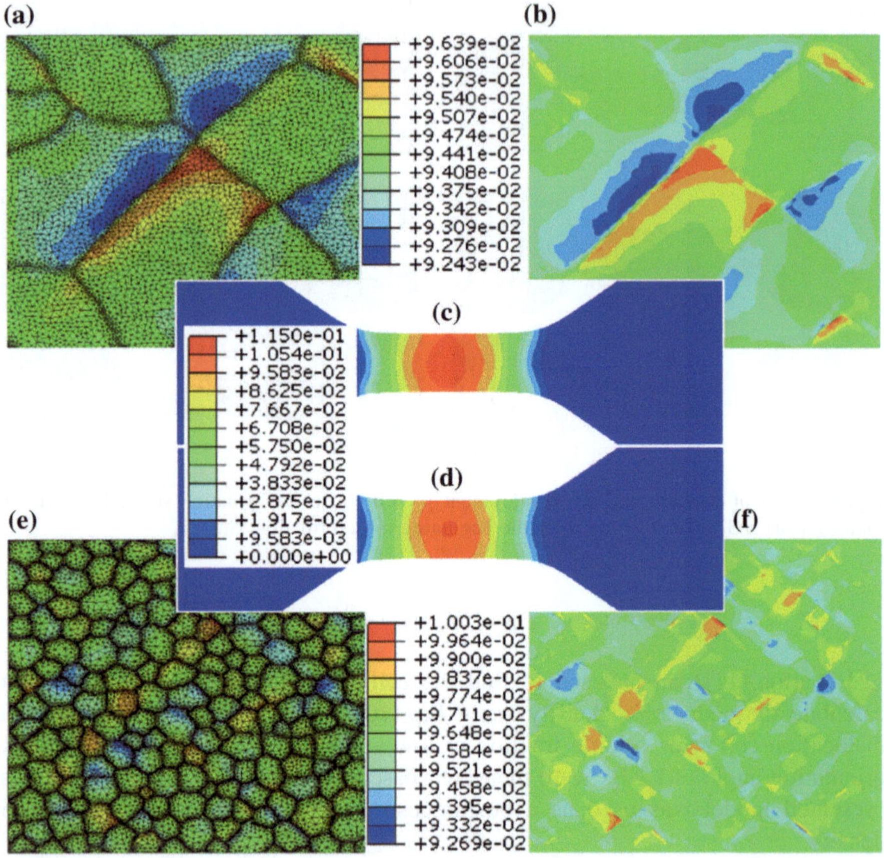

Fig. 7 Equivalent strain distribution in submodels of specimens subjected to deformation route of 2a (**a**) and 5 (**e**). Equivalent strain distribution in global model: route 2a (**c**), route 5 (**d**). Results with removed finite element mesh: route 2a (**b**) and 5 (**f**)

Fig. 8 Mises stress distribution in specimen from A steel subjected to the deformation route 2a—submodel

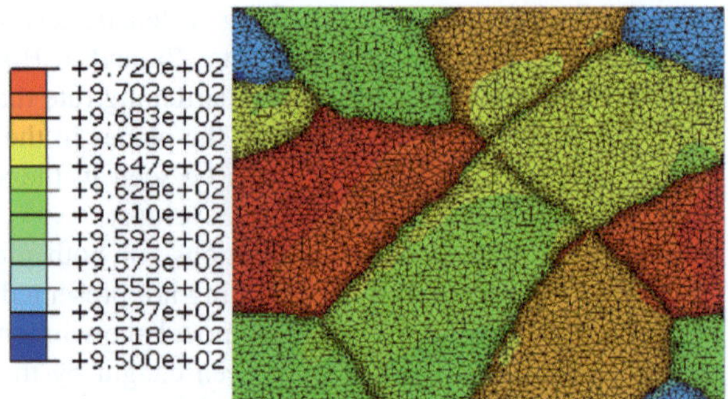

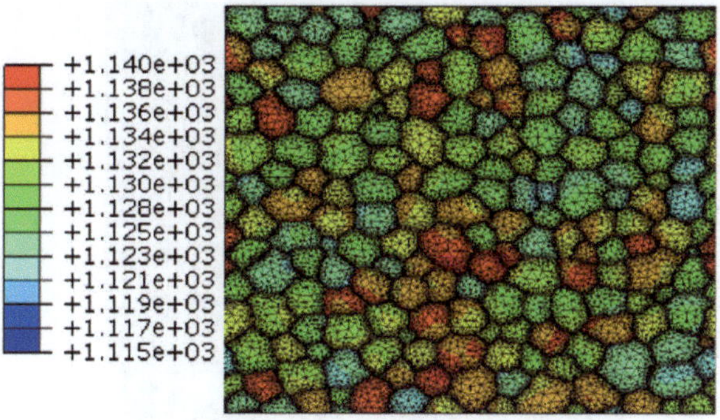

Fig. 9 Mises stress distribution in specimen from A steel subjected to the deformation route of 5—submodel

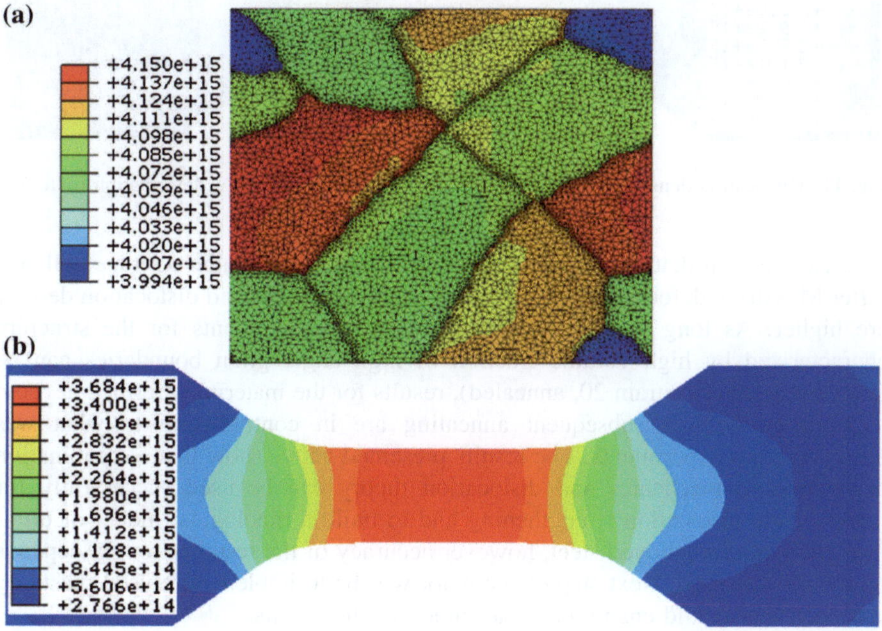

Fig. 10 Dislocation density in local (**a**) and global (**b**) model—steel A, deformation route 2a

strain localization bands are visible between the grains. These results describe material behavior more precisely than the global model (Fig. 7c and d).

Mises stress distributions are presented on Figs. 8 and 9 for the submodels deformed according to the deformation schedules of 2a and 5, at the moment corresponding to the onset of necking. Similarly, dislocation density calculations using Holt model are summarized in Figs. 10 and 11 for global models and submodels.

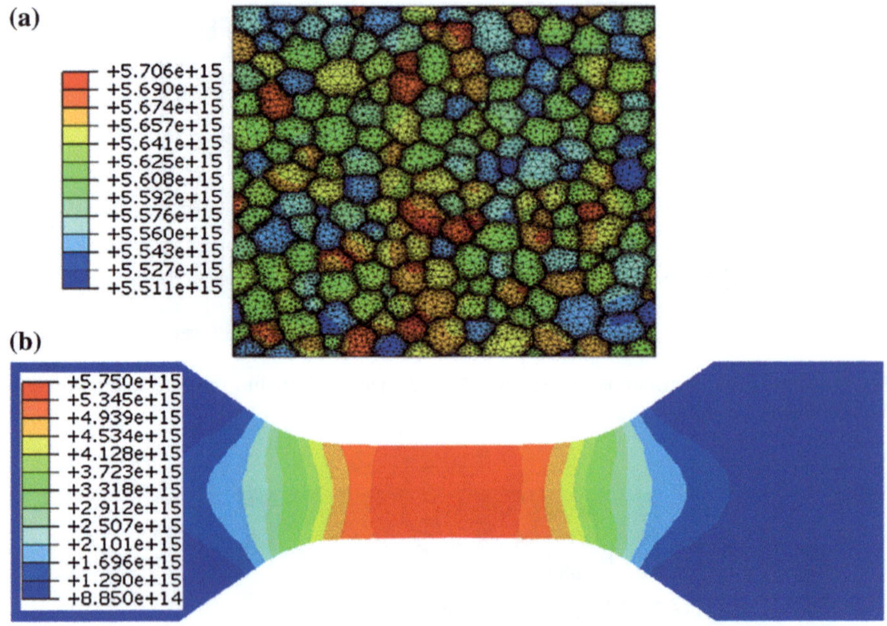

Fig. 11 Dislocation density in local (**a**) and global model (**b**)—steel A, deformation route 5

It can be seen that for the specimen characterised by smaller average cell size (after Maxstrain deformation route of 5), both Mises stress and dislocation density are higher. As long as the dislocation density measurements for the structure characterized by high volume fraction of high angle grain boundaries can be treated reasonably (strain 20, annealed), results for the material deformed at room temperature without subsequent annealing are in contradiction to what was observed from experiments. The results presented above show that a combination of continuum mechanics and dislocation theory can be used successfully for mapping the dislocation strengthening and to build a rheological model of ultra-fine-grained microalloyed steel, however accuracy of the results strongly depends on the models used. Next step of the work will be to implement crystal plasticity approach that would enable better accuracy of the results.

In Fig. 12 values of the rate of work hardening rate versus true strain calculated in global models using Abaqus Standard were plotted and compared with the experimental data. As it was already mentioned, basing on a true stress-true strain curve and using criteria of plastic instability, it is possible to determine, in unequivocal way, the field of the uniform elongation by Considére criterion. Uniform plastic deformation occurs as long as the true stress is below the value of work hardening rate ($d\sigma/d\varepsilon < \sigma$). When the two quantities are equal, the uniform deformation stops and necking begins. Comparing measured and calculated values of the uniform elongation it can be noticed that values calculated in Abaqus are slightly different from those, calculated based upon tensile tests.

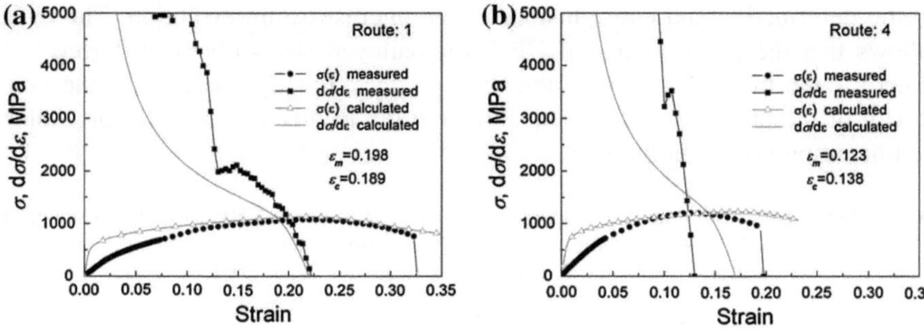

Fig. 12 Comparison of measured and calculated values of the uniform elongation for specimens subjected to deformation routes of 1 (**a**), and 4 (**b**). Steel A

5 Summary

Results of the computer simulations are compared with those obtained experimentally. The SPD processes were used to systematically evaluate the effects of very large strains on the micro-structure evolution and mechanical properties of microalloyed steel. The influence of the deformation conditions on the final microstructure and inhomogeneity were analyzed. Presented results show that there is obvious difference between the relative grain size distributions and homogeneity of the two investigated steels. It is clearly evidenced by the experimental results that presence of the microalloying elements in the steels subjected to very large strains supports processes of dislocation substructure formation and finally leads to ultra fine-grained microstructures. The scope of this work also confirmed the possibilities of using computer modeling to design, predicting and analyzing the processes taking place in the SPD and the effect of very large strain on the microstructure evolution and ductility of microalloyed steels although no clear methodology exists for transfering this knowledge into a proper macroscopic model. It was shown, that thanks to multiscale approach however, parameters such as equivalent strain and stress, or dislocation density at the moment corresponding to the onset of necking can be calculated at different scales, what makes the analysis of the mechanical behavior of UFG materials much more precise. As a result, many benefits can be expected. For example, a decrease in the product weight can be realized without the need of making compromises on mechanical properties, safety, or quality. Beyond this, using innovative steel products and production processes makes a reduction of manufacturing costs possible. The expected benefits are clear and hopeful. However, the fact that such modern materials as UFG microalloyed steels can be produced, does not mean that prediction of the behaviors of these materials in different conditions of their structural use environment is possible. These problems can be addressed when the complete, multiscale model is formulated. The strength and ductility of the modern microalloyed steels strongly depend on the strain path changes, accumulated energy of plastic deformation and entire history of the microstructure evolution. The thesis follows that the good quality of UFG microalloyed steels obtained during SPD directly confirms the quality of the final product. Finally, because of the complexity of the problems mentioned, further study in this area is necessary before significant progress can be made.

Acknowledgments The financial support of MNiSW (Grant no. N N508 3982 37) is gratefully acknowledged. Authors are grateful to Dr. Ł. Madej from AGH University of Science and Technology for generation FEM meshes of grained structures using DMR software.

References

1. Alexandrov, I.V.: Mater. Sci. Forum **584–586** (2008)
2. Belyakov, A., Sakai, T., Miura, H.: Mater. Trans. JIM **41** (2000)
3. Conrad, H.: Mater. Sci. Eng. A **341** (2003)
4. Considère, A.: Mèmoire sur l'emploi du fer et de l'acier dans les constructions, pp. 5–149. Paris Press, Paris (1886)
5. Copreaux, J., Lanteri, S., Schmitt, J.-H.: Mater. Sci. Eng. A **164** (1993)
6. De Borst, R.: Comput. Mater. Sci. **43** (2008)
7. Hall, E.O.: Proc. Phys. Soc. London Sect. B **64** (1951)
8. Hansen, N.: Scr. Mater. **51** (2004)
9. Hansen, N.: Mater. Sci. Eng. A **409** (2005)
10. Hirth, J.P., Lothe, J.: Theory of Dislocations, 2nd edn, pp. 5–857. Wiley, New York (1982)
11. Holt, D.L.: Appl. Phys. **41** (1970)
12. Howe, A.A.: Mater. Sci. Technol. **25** (2009)
13. Humphreys, F.J., Chan, F.M.: Mater. Sci. Technol. **12** (1996)
14. Jazaeri, H., Humphreys, F.J.: Acta Mater. **52** (2004)
15. Jia, D., Ramesh, K.T., Ma, E.: Acta. Mater. **51** (2003)
16. Khan, A.S., Huang, S.: Int. J. Plast. **8** (1992)
17. Khan, A.S., Suh, Y.S., Chen X., et al.: Int. J. Plast. **22** (2006)
18. Ko, Y.G., Shin, D.H., Park, K.-T., et al.: Scr. Mater. **54** (2006)
19. Kocks, U.F., Canova, G.R.: In: Hansen, N., Leffers, T., Lilholt, H. (eds.) Deformation of Polycrystals: Mechanisms and Microstructures, p. 185. Riso National Laboratory, Roskilde (1981)
20. Ma, E.: JOM **58**(4), 49–53 (2006)
21. Madej, Ł.: Development of the Modeling Strategy for the Strain Localization Simulation Based on the Digital Material Representation, pp. 124–140. AGH University Press, Krakow (2010)
22. Madej, Ł., Rauch, Ł., Yang, R.: Arch. Metall. Mater. Sci. **54** (2009)
23. Majta, J., Doniec, K., Muszka, K.: Mater. Sci. Forum **638–642** (2010)
24. Majta, J., Muszka, K.: Mater. Sci. Eng. A **464** (2007)
25. Majta, J., Pietrzyk, M., Lenard, J.G.: Mater. Sci. Eng. A **208** (1996)
26. Muszka, K., Doniec, K., Majta, J.: Comput. Meth. Mater. Sci. **9** (2009)
27. Muszka, K., Hodgson, P.D., Majta, J.: J. Mater. Process. Technol. **177** (2006)
28. Muszka, K., Hodgson, P.D.: J. Majta, Mater. Sci. Eng. A **500** (2009)
29. Muszka, K., Majta, J., Hodgson, P.D.: ISIJ Int. **47** (2007)
30. Nes, E., Marthinsen, K., Holmedal, B.: Mater. Sci. Technol. **20** (2004)
31. Oscarsson, A., Hutchinson, B., Nicol B., et al.: Mater. Sci. Forum **157–162** (1994)
32. Petch, N.J.: J. Iron Steel Inst. **174** (1953)

33. Sellars, C.M.: In: Rodriguez-Ibabe, J.M., Gutierrez, I., Lopez, B. (eds.) Modelling Strain Induced Precipitation of Niobium Carbonitride during Hot Rolling of Microalloyed Steel. Materials Science Forum, vol. 500–501, p. 73. TTP, Donostia-San Sebastian (1998)
34. da Silva, M.G., Ramesh, K.T.: Int J Plast 13 (1997)
35. Song, R., Ponge, D., Raabe, D., Speer, J.G., Matlock, D.K.: Mater. Sci. Eng. A **441** (2006)
36. Takayama, A., Yang, X., Miura, H. et al.: Mater. Sci. Eng A **478** (2008)
37. Taylor, G.I.: J. Jpn. Inst. Met. **62** (1938)

Mechanical Behaviour of Al 6061-T6 Aluminium Alloy Under Large Strain and Failure

M. Giglio, A. Gilioli and A. Manes

Abstract Severe and extreme loads, that introduce large strains and failure, are a present challenge in the design of critical mechanical components. Even though full scale testing is a fundamental approach for reliable structural integrity evaluation, numerical simulation is an alternative economical method that is now increasingly chosen especially because of the development of computing performances. In particular, the numerical assessment of the ductile fracture in metallic components represents an innovative and challenging field in the structural integrity scenario. Aerospace, automotive and manufacturing industries have recently boosted their interest in these kinds of simulations with the aim to make these approaches, little by little, reliable also for certifications. Taking the requirement to characterize material for further impact simulations as a starting point, the work described in this chapter contains a complete characterization of the mechanical properties of Al 6061-T6 aluminium alloy as far as material hardening and fracture locus are concerned. The calibration has been carried out through a series of experimental tests on simple specimens. These specimens have similar geometry, but are subjected to different stress triaxiality, thanks to the use of a multiaxial hydraulic test machine. All the experimental tests have been numerically simulated and a complete material constitutive model has been calibrated, which based on the results of these experimental analyses.

Keywords Fracture mechanism · Mechanical testing · Ballistic impact · Finite elements

M. Giglio (✉) · A. Gilioli · A. Manes
Dipartimento di Meccanica, Politecnico di Milano, Via la Masa 1, 20156 Milano, Italy
e-mail: marco.giglio@polimi.it

A. Gilioli
e-mail: andrea.gilioli@mail.polimi.it

A. Manes
e-mail: andrea.manes@polimi.it

N. Bonora and E. Brown (eds.), *Numerical Modeling of Materials Under Extreme Conditions*, Advanced Structured Materials 35, DOI: 10.1007/978-3-642-54258-9_7,

1 Introduction

Nowadays safety and reliability in extreme condition is one of the main issues addressed especially to aerospace industries. In this scenario the evaluation and the assessment of the residual structural integrity after damage is a key subject. The major advance in this direction aims at the development of damage tolerant analyses which are able to state the residual life of damaged components. However, these analyses need a reliable numerical description of the damage phase in order to carry out the next task.

Taking military aerospace components as an example, a projectile impact can be a critical and an extreme condition for the structural assessment. This is particularly true when military rotorcrafts, designed for low altitude mission in a hostile environment, are concerned. The evaluation of the behaviour of high critical components after the ballistic impact is crucial. Although experimental tests remain fundamental, at present, the numerical approach allows the simulation of such complex scenario. Once the impact conditions are known, the subsequent accurate description of the mechanical properties and behaviour of the materials involved is a fundamental step in the simulation of this phenomenon. However, two elements are required:

- a constitutive law which is able to describe the mechanical behaviour of the material (in particular the plastic flow in the hardening phase);
- a failure criterion to model the ductile failure of the material.

Ballistic impact implies that the target material is subjected to extreme load conditions in which a very high strain rate is combined with high levels of deformation. However, such extreme conditions can be also found in technological processes like trenching, moulding and other techniques and hence also the industry shows an interest in these kinds of simulations.

According to this introduction, the main purpose of this work is the complete calibration of the Al 6061-T6 aluminium alloy. This research is driven by the challenge to correctly describe the damage of a transmission shaft after a ballistic impact using a finite elements model. The following chapter describes, the complete methodology, based on a virtual test approach, used for the calibration of both constitutive law and failure criterion for the Al 6061-T6. This methodology uses extensive experimental tests and numerical support. Particular attention was paid to the failure calibration, because the calibration of a versatile fracture criterion which covers a large range of conditions requires a number of different experimental tests. In this work the Bao–Wierzbicki's approach [1] of the problem is used as a main framework. This approach is, however, expanded by the introduction of further new different experimental tests, such as pure torsion, biaxial torsion-traction and torsion-compression tests. All experimental tests have been numerically simulated with the commercial finite elements software ABAQUS, in order to carry out a complete "virtual test".

As just underlined the correct reproduction of the fracture is a key issue for the work presented, therefore a literature survey on this topic is hereafter presented. Bao and Wierzbicki analyse nine of the most frequently used criteria in the literature [1] and they demonstrate that none of them can fully describe the onset of failure in different kinds of tests such as upsetting tests and tensile tests (smooth, notched round specimen). The onset of the failure condition strongly depends on the stress state and in particular on the stress triaxiality. Most of the criteria described in literature are designed and calibrated with specific tests that cover a reduced range of stress triaxiality. Moreover the calibration of a ductile failure criterion is a time consuming process with a large number of parameters involved: for example the Gurson model [2, 3] takes into account ten different parameters. The main drawback of the Gurson model in many practical applications lies in the complexity of its calibration. However, the literature reports several models which predict the onset of ductile failure. Four frameworks are described: abrupt models, porous solid mechanics models, continuous damage mechanics models (CDM) and cohesive models.

Abrupt models (the framework that we also have adopted in this work) are increasingly used in industrial application due to their relatively easy calibration and their diffusion in the commercial finite element code. The criteria are based on the assumption that fracture happens when the cumulative plastic strain reaches a critical value, see Eq. (1):

$$\int_0^{\overline{\varepsilon_{pl}}} f(stress\,states, strain\,rates, temperature, \ldots)\, d\varepsilon_{pl} \geq D_c(material) \tag{1}$$

where f is a weight function that depends on a number of parameters. Examples of these criteria are McClintock [4], Cockcroft and Latham [5], Johnson and Cook [6]. Although all of these criteria are relative simple, they provide good results only in specific load conditions. In [1] Bao and Wierzbicki propose a better approach to the problem. The weight function "f" can be called "fracture locus" and it depends mainly on the stress triaxiality.

Because the mechanisms of crack formation are very different at various triaxialities, Bao and Wierzbicki define a fracture locus which is not continuous. Instead they define the fracture locus with different equations for a different range of stress triaxiality. The great influence of stress triaxiality on ductile damage of metals has been also highlighted in [7]. Additionally they point out the necessity to develop a damage criterion able to describe different failure modes depending on triaxiality. Recent studies [8] have demonstrated that strain at failure depends not only on triaxiality but also on a deviatoric parameter of the stress state and in particular on the Lode angle.

A different way to analyse damage and failure problems is the continuous damage mechanics (CDM) approach. This method assumes that processes responsible for macroscopic failure are strictly related to the material microstructure. CDM models evaluate the damage accordingly by analysing the behaviour of the affected material at a macro scale level. In particular damage is considered as one of the thermodynamics state variables which take into account the progressive loss of material load carrying which in turn results from irreversible modifications of the microstructure. One of the most important and recent CDM model is the Bonora's one [9]. Another important CDM model has been developed by Lemaitre [10]. In Particular in recent years Bonora's model has been applied and improved. In [11] authors demonstrate the geometrical transferability of damage parameters of Bonora's model to geometry different to the one exploited for their calibration. Good results have been found in FE simulations of notched bar and also for cracked geometry (high triaxiality). The tested material was an A533 grade B steel. The geometrical transferability has been also verified in [12] using flat notched specimen. In [12] an experimental procedure has been described in order to evaluate the four material parameters of Bonora's model. In [13] authors showed that Poisson's ratio doesn't affect damage for a ferritic high alloy steel, a carbon manganese steel vanadium microalloyed and a high purity 99.9 % copper.

Although the CDM approach is promising and effective, in this chapter, authors have decided to focus on the calibration of an abrupt model because of its simplicity, the availability of its commercial code and its capability to reliably describe a different stress state (triaxiality).

Moreover, the final part of the present chapter has been dedicated to the validation of the material calibration through ballistic impact simulations where the stress triaxiality range is relatively wide.

2 Al 6061-T6

Generally aluminium alloys have good mechanical properties, high corrosion strength and low density. Nowadays these alloys are widely used in many engineering fields and in particular in contexts in which weight reduction is a critical factor such as in aerospace applications. Although pure aluminium is one of the most widespread elements on earth, it is too soft and ductile. Consequently it is often combined with many different elements, such as copper (Cu), silicon (Si), magnesium (Mg), zinc (Zn), manganese (Mn) to form alloys. Al 6061-T6 belongs to the aluminium series 6,000 with silicon and magnesium as principal alloying components. These alloys are formed by heat treatment (in present case T6) and they show good formability, workability, weldability and shaving. The chemical composition of Al 6061-T6 is described in Table 1.

In Table 2 the main physical-chemical properties of Al 6061-T6 are summarized [14].

The T6 suffix describes the heat treatment applied to the alloy, T6 standing for a two phase treatment. The entire process is based upon the precipitation hardening phenomenon and results in a large strength increment with only a minimal

Table 1 Chemical composition of Al 6061-T6

Al (%)	Mg (%)	Si (%)	Fe (%)	Cu (%)	Mn (%)	Cr (%)	Zn (%)	Ti (%)
98	0.8–1.2	0.4–0.8	0.7	0.15–0.4	0.15	0.04–0.35	0.25	0.15

Table 2 Physical–chemical properties of Al 6061-T6

Volumic mass	2,700 (Kg/m^3)
Specific heat	0.89 (J/Kg°K)
Sound velocity	5,350 (m/s)
Elastic modulus	70,000 (MPa)
Shear modulus	26,000 (MPa)
Poisson's coefficient	0.33

reduction of the break elongation. In the first phase of the treatment the alloy is warmed up to 495–505 °C for about 30 min followed by natural cooling. It is then followed by the second phase, also called the aging phase, in which the alloy is re-heated to a temperature of 160 °C and kept at that temperature for 18 h.

3 The Material Constitutive Law: Johnson–Cook

In order to build a widely applicable numerical model which is able to provide good results, one of the key features is the calibration of the constitutive law of the material. This aspect is crucial especially for numerical models which must describe complex and severe conditions in a plastic range, like the phenomena which occur during ballistic impact. The Johnson–Cook (JC) law is a common and versatile constitutive law. It offers many advantages, most importantly: its simple calibration made with few parameters in order to reproduce strain-rate and temperature effects, its ease to be implemented into many types of FE software and the availability of its parameters in literature. The JC law is a semi-empiric law which describes in one single equation the influence of both temperature and strain rate on the stress-strain curve. The complete equation of the JC constitutive law is the following Eq. (2):

$$\sigma = \left[A + B(\varepsilon_p)^n\right]\left[1 + C\ln\left(\frac{\dot{\varepsilon}_p}{\dot{\varepsilon}_0}\right)\right]\left[1 - \left(\frac{T - T_a}{T_f - T_a}\right)^m\right] \tag{2}$$

- A is the material elastic limit;
- B and n are parameters which describe the plastic behaviour of the material and its hardening;
- ε_p is the plastic strain;
- C is a parameter which describes the influence of the strain rate;
- $\dot{\varepsilon}_p$ is the plastic strain rate;

Table 3 Different values of the JC's parameter C

C	Reference
0.012	[34]
0.011	[22]
0.010	[35]

- $\dot{\varepsilon}_0$ is a reference strain rate;
- T is the material temperature;
- T_a is the room temperature at which the tests have been done;
- T_f is the melting temperature;
- m is a material characteristic parameter.

As previously mentioned, JC is a semi-empiric model which means that is based upon the interpolation of experimental data with a curve which has a specific mathematical form and in which some parameters are fitted to obtain a final curve able to fit the numerical data well. The fundamental hypothesis of JC model is that is possible to uncouple the effect of strain rate and temperature. This means that the JC criterion is a common power law multiplied for appropriate expression able to take into account the effect of strain rate and temperature. Thus the base expression for JC can be achieved by common tensile tests and the effects of strain rate and temperature can be added (both or for only one of them) with data acquired from more complex tests.

Strain rate has a very important influence on a ballistic impact due to the high velocity of the bodies involved in such phenomena. The velocity of the bullet is about 850 m/s for the ballistic impact simulated in this chapter, and this means that the expected strain rate is beyond $10^3\ s^{-1}$. In presence of such elevated strain rate, the behaviour of the material can drastically change and hence the strain rate influence can't be neglected. In the JC model this effect is included in the C parameter. Al 6061-T6 specific values for this parameter have been reported in the literature [8–10]. Three different references state very similar values and thus a mean value of the three have been chosen for this publication (Table 3).

Recent papers regard the temperature effect on the constitutive law of Al 6061-T6. In particular the work of Colombo and Giglio [15] shows that the temperature effect can be neglected for the impact conditions considered in the present chapter.

It is interesting to point out that the JC model isn't the only model able to describe metallic mechanical behaviour (constitutive law) under extreme conditions. A different approach based on the real physics of the problem is the mechanical threshold stress model (MTS). This model has been developed by Follansbee and Kocks [16] for OFHC copper and Goto et al. for HY-100 steel [17]. A physical approach means that the MTS model tries to describe the real dislocation motion responsible for the flow stress by taking into account an internal state variable called mechanical threshold stress. The MTS model can furthermore take the temperature and the strain rate effect into account. To achieve this, the [18] MTS plasticity model is combined with a nucleation and growth model

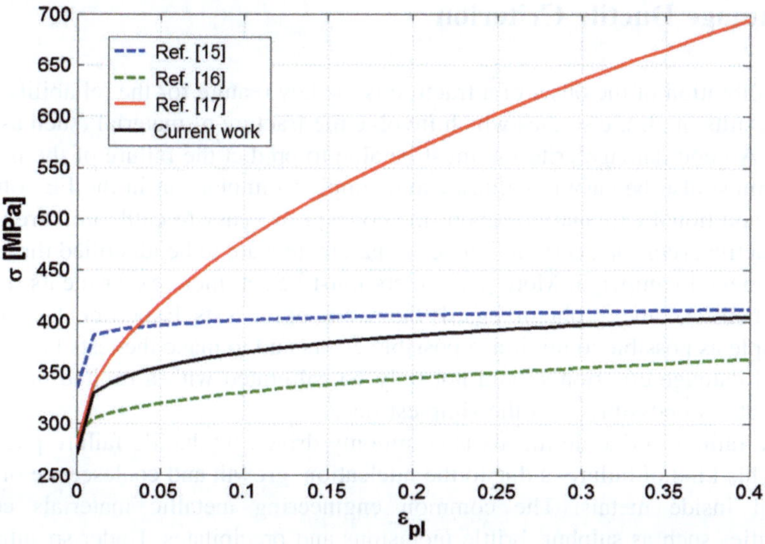

Fig. 1 Comparison between the JC curves for Al 6061-T6 calibrated in the present work and in the published literature

Table 4 Different literature parameters for JC constitutive law for Al 6061-T6

A (MPa)	B (MPa)	n	C	$\dot{\varepsilon}_0$ (s^{-1})	m	State	References
335.0	85.0	0.110	0.012	1	1.00	Forged	[34]
289.6	108.0	0.420	0.011	1	1.34	Laminated	[22]
280.0	673.0	0.530	0.010	1	n.d	n.d.	[35]
270.0	**165.5**	**0.222**	**0.011**	**1**	**n.d**	**Laminated**	**Present work**

Data from present work in bold

(NAG) based on microscopic assumptions in order to reproduce failure during a planar impact test. Although a physical approach has various interesting aspects, a phenomenological approach has been chosen instead in this chapter. The main reason for this choice is the availability of data (both experimentally and also in the literature) for the JC model, which essentially renders a phenomenological approach much more interesting for real industrial applications. Additionally a phenomenological approach matches very well with an abrupt ductile criteria framework like the one followed in this chapter. Thus the JC model has been calibrated. The methodology used to calibrate the constitutive law will be fully explained in following paragraphs because it is deeply related at a virtual test approach. Figure 1 and Table 4 in the current paragraph, however, provide an interesting comparison between the JC constitutive law calibrated in the present chapter and other JC models from the literature.

4 Damage Ductile Criterion

The calibration of the onset of a fracture is the key feature for the reliability of the final results in all the studies which involve the fracture of material (such as crash tests). A good damage criterion must be able to predict the failure of the material but it must also be easy to calibrate and simple to implement in the FE software. Some additional explanations about the concept of "easy to calibrate" are useful. In a ductile criterion a certain number of parameters are to be identified through an experimental campaign. More parameters must be set, more experiments must be carried out. Another problem regards the nature of the tests. Experiments should be as simple as possible to minimize possible errors and to make the tests faster. Thus a good damage criterion should not only be calibrated with a limited number of tests but also possibly with the simplest ones.

The failure in ductile metals is commonly driven by ductile failure phenomenon. This kind of failure is due to the nucleation, growth and coalescence of voids present inside metals. The common engineering metallic materials contain impurities such as sulphur, brittle inclusions and precipitates. Under specific load conditions, the metal matrix unglues from the impurity and forms micro voids that can grow until they join each other (coalescence) and form micro cracks. These micro cracks can grow and reach a macroscopic dimension that can induce a slant fracture. Numerically speaking ductile abrupt criteria are based upon the definition of a damage parameter D:

$$D = \int_{0}^{\overline{\varepsilon_f}} \frac{d\varepsilon_{pl}}{\varepsilon_f\left(\eta, \dot{\varepsilon}_{pl}, T\right)} \tag{3}$$

Accordingly to a finite approach, the parameter D can be calculated as the summation of the ratio between plastic strain ε_{pl} and plastic strain at failure ε_f. The plastic strain at failure ε_f is not a fixed value but instead is a function of stress triaxiality η, plastic strain rate $\dot{\varepsilon_{pl}}$ and temperature T. The most influencing parameter is the stress triaxiality η which is defined in (4):

$$\eta = \frac{\sigma_h}{\sigma_{VM}} \tag{4}$$

where σ_h is the hydrostatic stress and σ_{VM} is von Mises equivalent stress. Hence it is very common to neglect the effect of the plastic strain rate and the temperature (see next paragraph) and to take into account only the stress triaxiality dependence. The curve which describes the relation between ε_f and η is commonly named fracture locus. The working mechanism of damage abrupt criterion implemented into numerical FE software is based on the onset of the damage initiation. When the parameter D reaches the unity value, the failure takes place. The elements completely lose their stiffness, are unable to support any additional loads and are thus eliminated from the analysis (stiffness reduced to zero).

4.1 Johnson–Cook Fracture Locus

One of the most common fracture locus frameworks is the Johnson–Cook fracture locus elaborated by the same authors of the constitutive law. The JC fracture locus considers the effect of stress triaxiality, temperature and strain rate:

$$\varepsilon_f = \left[D_1 + D_2 \exp(D_3 \frac{\sigma_h}{\sigma_{VM}})\right]\left[1 + D_4 \ln\left(\frac{\dot{\varepsilon}_{pl}}{\dot{\varepsilon}_0}\right)\right]\left[1 + D_5 \frac{T - T_0}{T_m - T_0}\right] \quad (5)$$

where D_1, …, D_5 are experimental constants. However, some problems arise from the direct use of the JC fracture locus. The ε_f limit curve estimates a monotonic decrease with the incrementation of η. For high stress triaxiality this curve can also show negative values with an inconsistent behaviour. Another problem of this curve appears for a negative stress triaxiality $\eta = -1/3$. Experiments have shown that for this specific value, plastic strain at failure ε_f grows to an infinity value but this evidence is not considered by the JC fracture locus law. Despite of these problems the main advantage of this fracture locus is that it is very powerful and easy to implement in the FE software. In the present chapter the effect of strain rate and temperature on the Al 6061-T6 fracture locus has been neglected. This means that the fracture locus is considered to depend only on the stress triaxiality. In the literature, Clausen et al. [19], Børvik et al. [20], Johnson and Holmquist [21] have shown that there is only a slight increment of ε_f increasing the strain rate. In Fig. 2 a JC fracture locus for Al 6061-T6, obtained from the literature data [22], Table 5, can be seen. In particular the effect of strain rate is highlighted. The maximum difference between ε_f at different strain rates is about 12 % for a strain rate of 10^5 s^{-1}. The dependence of ε_f from the strain rate is clearly largely negligible in comparison with the dependence from the stress triaxiality η. Furthermore also the effect of temperature on the fracture locus has been neglected in the current chapter. Figure 3 shows a JC fracture locus for Al 6061-T6 from the literature [22] with its dependence on temperature. The effect of temperature on ε_f starts to become sensitive for $T \geq 200$ °C; below this temperature, the effect is negligible. Zukas [23], Mayer [24], Colombo and Giglio [15] show that, in conditions very similar to the ones applied in the present work for ballistic impact, the maximum temperature increment for Al 6061-T6 is about 130 °C. In the case of the problem investigated it is thus legitimate to neglect the temperature effect on the fracture locus. It is, however, very important to point out that effects of the strain rate and temperature can be only neglected for the fracture locus. In fact the most important parameter on which ε_f depends is the stress triaxiality while the temperature and the strain rate have a minor influence. Instead, as already stated in the previous paragraph, the effect of strain rate on the constitutive law cannot be ignored.

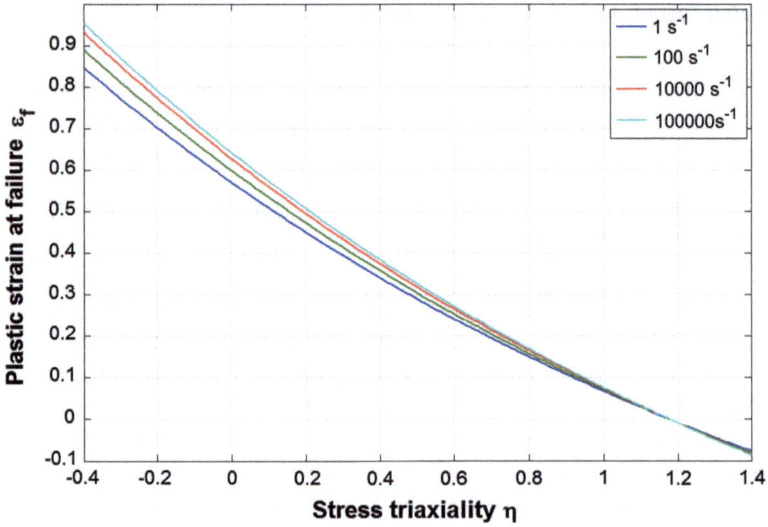

Fig. 2 Effect of strain rate on a literature JC fracture locus for Al 6061-T6 [22]

Table 5 Parameters of a literature JC fracture locus for Al 6061-T6 [22]

D_1	D_2	D_3	D_4	D_5	T_m (°K)	T_0 (°K)	$\dot{\varepsilon}_0$ (s^{-1})
−0.77	1.45	−0.47	0.011	1.6	925	294	1

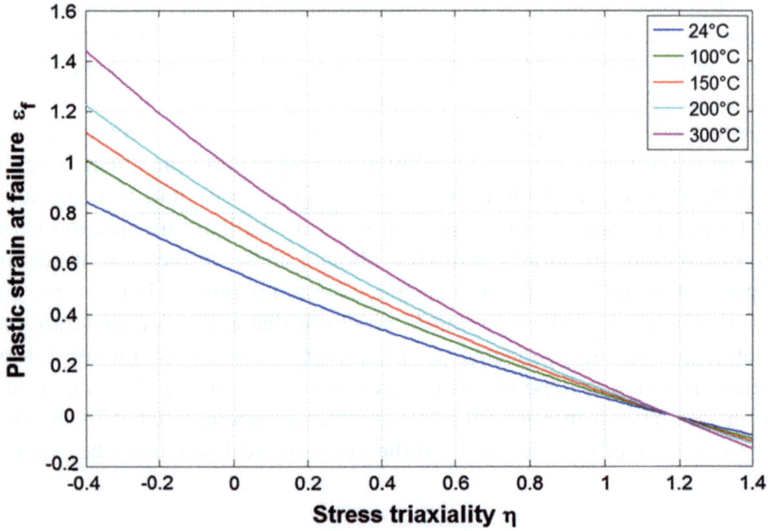

Fig. 3 Effect of temperature on a literature JC fracture locus for Al 6061-T6 [22]

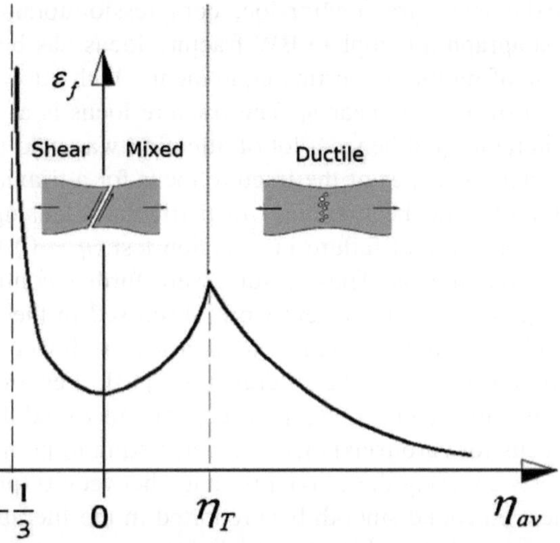

Fig. 4 Bao-Wierzbicki [1] fracture locus

4.2 Bao-Wierzbicki Fracture Locus

The exact prediction of the stress triaxiality values involved in a complex load case is difficult to achieve. This means that a fracture locus able to cover a large range of stress triaxiality is needed. Unfortunately the JC fracture locus covers only a rather narrow range of stress triaxiality because it is generally calibrated only with a tensile test on smooth and notched specimen, and is thus related to a 0.33–1 stress triaxiality range. In [1, 25] Bao and Wierzbicki (BW) develop a more flexible fracture locus. The authors point out that a simple continuous curve like the JC one doesn't correctly reproduce the material failure condition, especially for negative and low stress triaxialities, as well as for very high stress triaxialities. Consequently they divided the fracture locus into three different parts:

- high stress triaxialities $\eta > 0.4$
- low stress triaxialities $0 \leq \eta \leq 0.4$
- negative triaxialities $-1/3 < \eta < 0$.

Each part of the curve describes a phenomenological different type of failure. For high stress triaxialities the failure is due to voids which form, grow and coalesce. The failure mechanism for negative stress triaxialities, which means that the stress state consists mainly of compression, is not yet fully understood. This failure mechanism is called "shear decohesion" [1]. Instead for low stress triaxialities a mixed failure mode between void coalescence and shear failure exists, Fig. 4.

BW fracture locus is, however, more complex than a common JC curve and needs several experimental tests in order to calibrate the entire curve. In addition to the common tensile tests on smooth round specimen, also notched and multi axial tests (traction\torsion, compression\torsion) are needed. In the following paragraph a complete BW fracture locus has been calibrated with the innovative use of multiaxial testing equipment which allows the detailed investigation of the transition zone near η_t. The fracture locus is at present matter of investigation in literature and hence a lot of attention was paid to the low triaxiality test. Slightly different shapes of the fracture locus for a triaxiality range around zero have been reported in the literature. In particular Clausing [26], McClintock [27] found a lower strain at failure in a torsion test ($\eta = 0$) than in a tensile test ($\eta = 0.4$) in various metals. These results were further confirmed by the work of Wierzbicki [1]. Also in this present work, discussed in the following paragraph, the strain at failure, evaluated with a torsion test, is lower than in the tensile test. Different conclusion have been drawn in [13]. Ferritic steel, carbon manganese steel vanadium and 99.9 % pure copper fail to exhibit a minimum (of what) fracture locus for zero triaxiality. It is interesting to point out that there aren't any specific tests covering the triaxiality range between 0 (pure torsion) and about 0.4 (tensile test on round smooth bar) reported in the literature. Therefore this zone has been investigated in this present chapter using a multi axial machine.

5 Experimental Tests

In order to calibrate a fracture locus over different stress triaxiality values, several experimental tests are needed. In addition to the mono axial tests reported in the literature [1, 28–30], also multi axial tests have been carried out in present chapter. In detail, monoaxial tensile tests have been performed with three different kinds of specimen: smooth round and two differently notched round specimens. Two different kind of multiaxial tests on the other hand have been conducted: pure torsion and mixed tests (torsion plus tension, torsion plus compression). Some upsetting tests with cylindrical specimen with different height to diameter ratios (h/D) have been also performed. All specimens have been obtained from the same Al 6061-T6 aluminium alloy plate. The tests were repeated, at least two times, to confirm the reliability of the values obtained. However, in the following figures only one experimental curve is reported for each type of tests. All experimental tests cover a wide range of stress triaxiality from $-1/3$ up to 0.4. All tests that have been carried out are summarized in Table 6, whereas Fig. 5 describes the geometry of the specimen used in the experimental tests.

5.1 Experimental Test Apparatus

The mono axial tests have been done using an electro-mechanical MTS Alliance RF150 machine. The load data have been collected from a 150 KN load cell and the displacements have been acquired with an extensometer with a measurement

Table 6 Experimental tests details

Load case	Load type	Specimen type	Superimposed constant load
Monoaxial	Tension	Round smooth	–
Monoaxial	Tension	Round notched. Radius 12 mm	–
Monoaxial	Tension	Round notched. Radius 5 mm	–
Monoaxial	Torsion	Round smooth	–
Multiaxial	Torsion	Round smooth	Traction (4 KN)
Multiaxial	Torsion	Round smooth	Traction (8 KN)
Multiaxial	Torsion	Round smooth	Compression (12 KN)
Monoaxial	Compression	Cylinder	–

base of 25 mm. The extensometer renders the measures independent of stiffness of the machine, Fig. 6c. A servo hydraulic multi axial machine MTS 809 has been used for the multiaxial tests, Fig. 6a, b. The axial load has been acquired with a 250 KN load cell, while 2,200 KN m load cell has been used to acquire the torque. The angular rotation has been acquired by the clamp heads of the machine. Thus also the effects of the torsional stiffness of the machine have been included. The multiaxial tests have been done in two steps. Firstly by the application of a longitudinal load (tension or compression) and then, secondly, by imposing a rotation with a fixed rotational velocity. The rotational movement is continuous until specimen failure is reached, Fig. 6a, b.

5.2 *Monoaxial Tests for High Stress Triaxialities (>0.4)*

As previously remarked, in order to assess a reliable and complete fracture behaviour for different conditions (triaxiality level), various tests are needed. However, in the literature, the most frequently used criteria (Johnson–Cook) is calibrated with only one kind of test (>0.4) [20, 31, 32]. During the tests the "necking phase" that is reached when the plastic strains concentrate in a specific area of the specimen is clearly visible. In this area a crack starts and quickly propagates causing the failure of the specimen with a sudden fall of the load in the force-displacement curves, Fig. 9. Figure 7a, b, c show various specimen at the end of the experiments.

5.3 *Mono and Multiaxial Tests for Low Stress Triaxiality ($0 \le \eta \le 0.4$)*

Various tests have been carried out to investigate the failure mechanics for low stress triaxialities. As already mentioned above, reports in the literature usually only calibrate the fracture locus curves with tensile tests on round smooth and notched

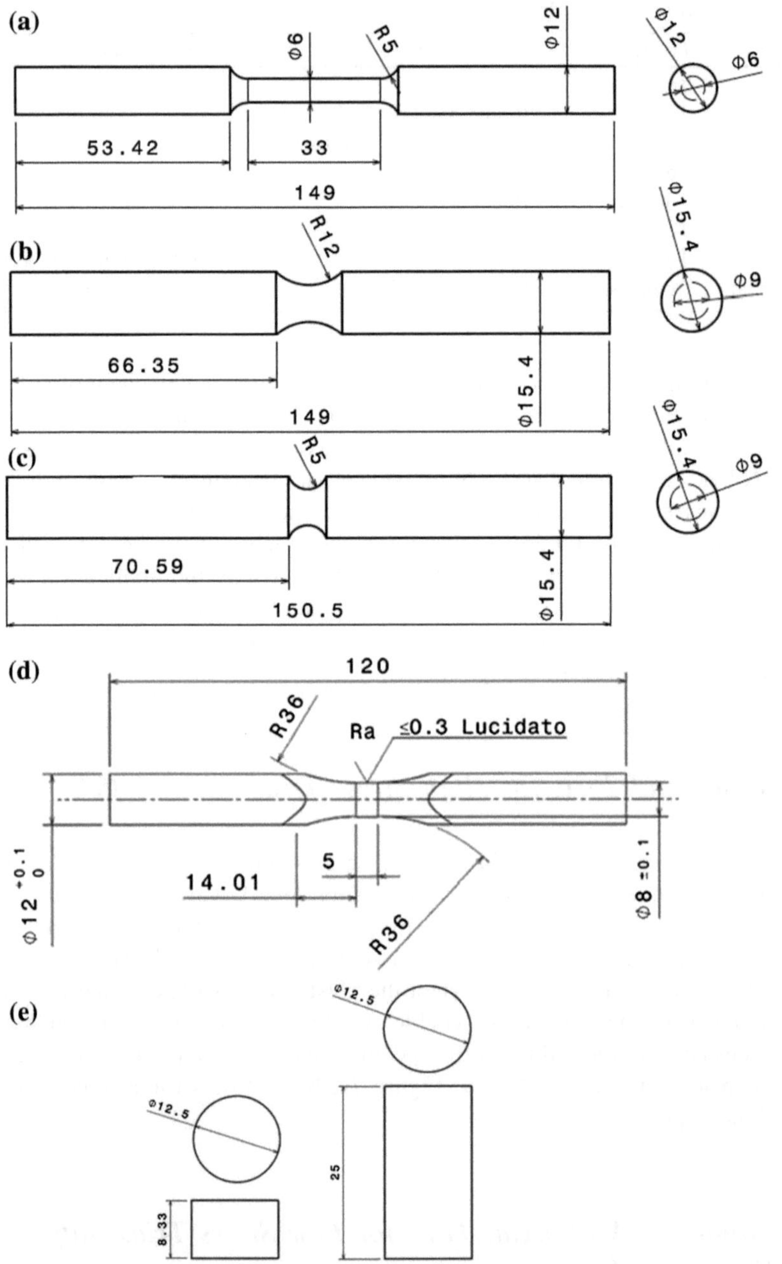

Fig. 5 **a** *Round* smooth specimen, **b** *round* notched specimen (notch radius 12 mm), **c** *round* notched specimen (notch radius 5 mm), **d** pure tension and multiaxial specimen, **e** upsetting specimens

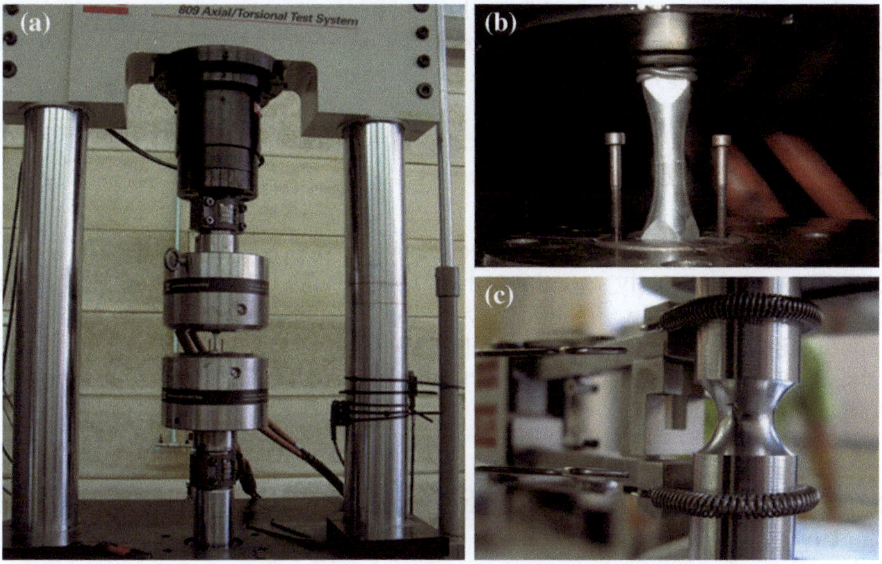

Fig. 6 **a** Multiaxial test machine, **b** pure torsion experiment, **c** tensile test with a notched specimen

specimen. This means that the fracture locus is extrapolated for stress triaxialities lower then about 0.3–0.4. For example Johnson and Cook in [6] carried out some pure torsion tests but these data was ignored and a continuous decreasing fracture locus for the entire range of stress triaxialities was assumed. In [27] McClintock studied the equivalent strain at failure ε_f of seven different materials in tension and torsion tests. He found that six of the tested materials showed a higher ε_f in tension than in torsion. Thus tests with a multi axial approach (a combination of torsion plus longitudinal, tension or compression, load) have been carried out. All tests were done on a servo hydraulic multi axial machine MTS 809 following the procedure explained above. Once the longitudinal load was applied, the rotational movement continued until the specimen rupture was reached. During the pure torsion and torsion/compression tests, no necking phenomena were observed. Instead during the torsion-tension test necking clearly appeared. However, the tensile test (high stress triaxiality) exhibited a different failure mode than the torsion and multi axial test (low stress triaxiality). In the high stress triaxiality experiment, the crack started after a necking phase and then the crack quickly propagated resulting in a sudden rupture. Instead in the low stress triaxiality tests the necking phase was evident only in the torsion-traction experiments. In these cases the propagation phase of the crack was very slow. No sudden drop of the torque was observable in the torque-rotation curve, however, the decrease of the load was noticeably smoother than in the tensile test curve. This indicates that the critical rotational value decreased with the increase of the tension load (from 4 to 8 KN). The critical rotational value is the critical angle at which the crack starts and where a sudden drop of the torque-rotation curve occurs. With a compression load, instead, the rotation to failure raises to very high values. In the torsion-compression test in fact it was impossible to reach fracture despite a rotation of 270°. Figure 7d, e, f show the specimen morphology at the end of the experiments.

5.4 Monoaxial Upsetting Tests

Some upsetting tests on cylindrical specimens with different h/d ratios were carried out. These tests mainly examined the experimental evidence that for a pure compression load case, fracture never appears. This scenario has also been reported in literature, for example in [33]. Also in the present chapter the specimens show a similar behaviour. No further numerical analyses have been performed on these tests.

6 Numerical Simulations of the Experimental Tests (Virtual Testing)

Virtual testing is a method used to perform a numerical reproduction of experimental tests (the most similar as possible), in this case the tests have been carried out in order to acquire the mechanical behaviour of materials. The most common comparison between numerical and experimental results in material specimens testing, concerns the load-displacement (rotation) curves. Hence one of the main purposes of virtual testing is to reproduce these curves with a numerical model in order to set the parameters of a material constitutive law. It then becomes possible to fit a correct hardening law and fracture locus. The accuracy of the parameters of the material law allows the exact simulation of complex phenomena like necking and ductile failure of the specimens. Obviously the full true stress-true strain curve is needed and an interactive approach is used to find the right parameters. Such a "trial and error" approach has been followed in order to fit the results obtained from the model with the experimental data.

An interesting alternative way which takes into account the necking effect on a true-stress curve can be found in [28]. This approach is based on the use of a special necking model called MLR and this technique permits the evaluation of the complete stress distribution and the plastic strain in a neck section of an unnotched specimen.

The numerical models have been analysed with the commercial finite element code ABAQUS V6.8. All the analyses have been conducted with an explicit code with quasi static approach, due to the high nonlinear behaviour. The elements used for the mesh construction have been of solid type with eight nodes and a reduced integration scheme (C3D8R). The average side dimension of the elements is 0.695 mm with 0.25 mm being the smallest dimension. In the central zone where the crack started, elements were of square shape with a side size equal to 0.65 mm. A compromised constitutive law has been obtained using a mean square approach

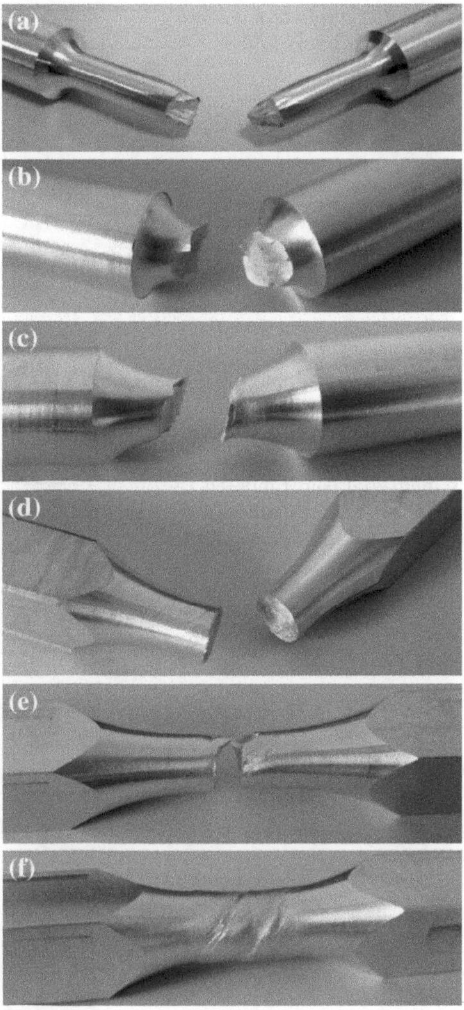

Fig. 7 Specimen at the end of the test. **a** *Round* smooth, **b** notched with 5 mm radius, **c** notched with 12 mm radius, **d** pure torsion, **e** torsion-tension, **f** torsion-compression

of the most optimized laws. In Table 7 the final constitutive law obtained for Al 6061-T6 has been reported and applied to all "virtual test" models. In Fig. 8 different numerical models just before the crack initiation can be observed. Each figure shows the value of the numerical variable DUCTCRT which represents the value of the damage D (red colour represents a value near to 1). When the parameter D reaches the critical value of 1, the corresponding element has its stiffness set to zero and it is eliminated from the analysis. The applied constitutive law grants a very good reproduction of the load-displacement and torque-rotation curves until the onset of fracture. Thus the virtual models are able to provide solid data for the subsequent reproduction of the fracture locus. The excellent match of numerical and experimental curves can be seen in Fig. 9.

Table 7 Constitutive law for Al 6061-T6 found in the present work

$\sigma = A + B(\varepsilon_p)^n$	
A	270 MPa
B	154.3 MPa
n	0.2215

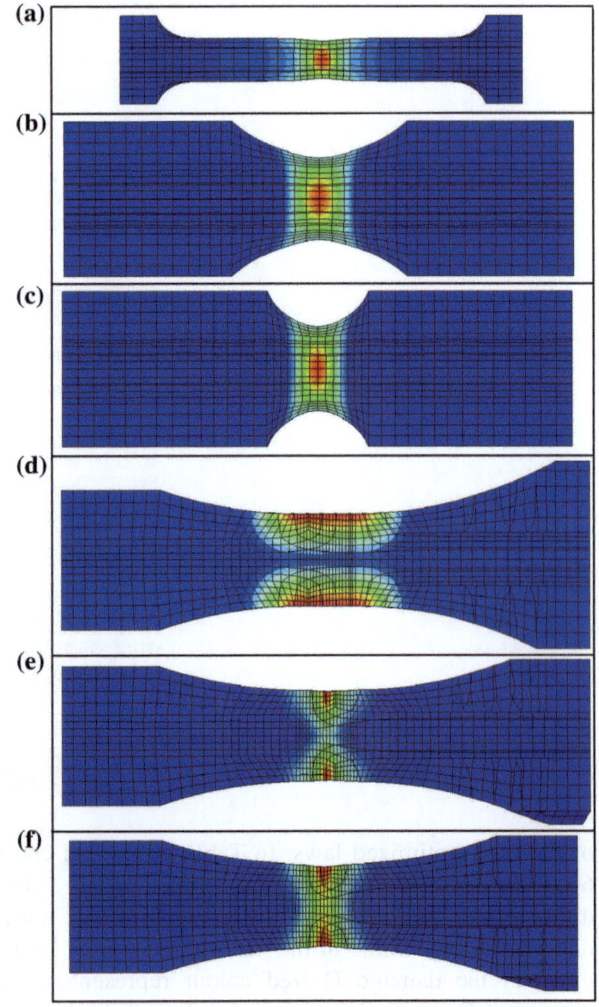

Fig. 8 Virtual test examples. *Colours* represent the value of parameter DUCTCRT (accumulated damage). *Red* is the highest value and *blue* is the lowest. **a** *Round* smooth, **b** *round* notched (notch radius 12 mm), **c** *round* notched (notch radius 5 mm), **d** pure torsion test, **e** multiaxial (superimposed tension 4 KN), **f** multiaxial (superimposed tension 8 KN)

The constitutive law chosen is compared with the laws in the literature in Fig. 1. It is interesting to note that among three curves in the literature, two are similar and one is definitely different from the other two. The constitutive law found in the present work lies in the middle between the two similar previously reported curves. Figure 9 shows the comparison between the load-displacement

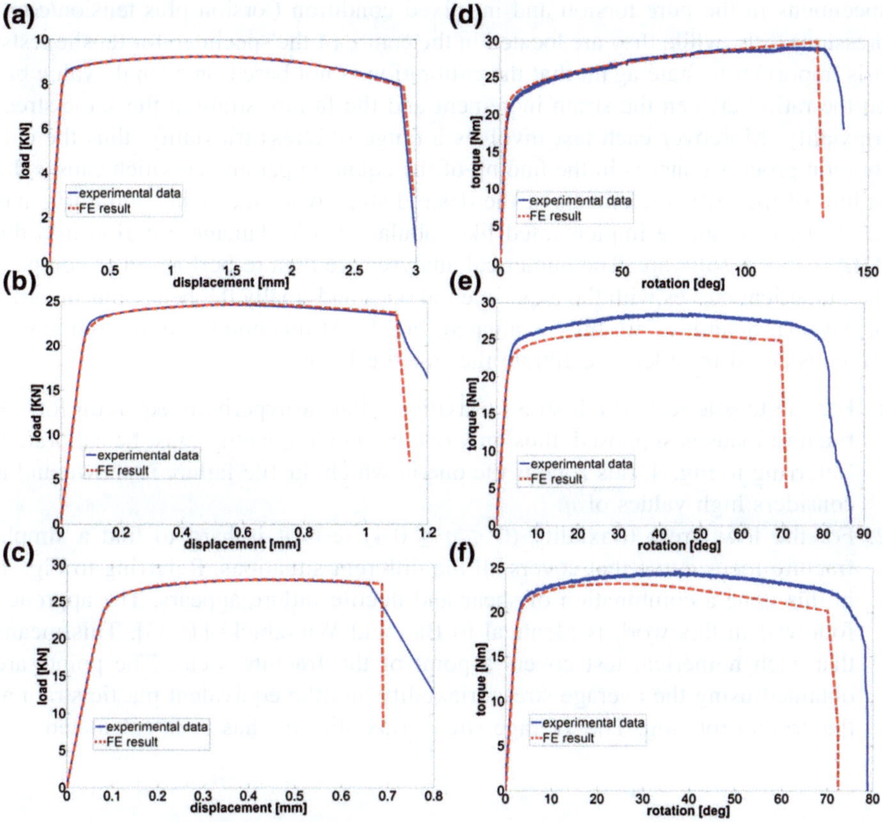

Fig. 9 Comparison between numerical and experimental load-displacement/torque-rotation curve for **a** *round* smooth, **b** *round* notched (notch radius 12 mm), **c** *round* notched (notch radius 5 mm), **d** pure torsion test, **e** multiaxial (superimposed traction 4KN), **f** multiaxial (superimposed tension 8 KN)

curves obtained experimentally and by the use of the corresponding optimized constitutive law. At this stage no damage criterion has been used and, therefore, no damage is present on the curves (on the contrary, in Fig. 9 the full calibration, including damage is presented).

7 Calibration of Fracture Locus

The results of the numerical analysis for the hardening fitting provide the starting point for the calibration of the fracture locus. Once the flow stress is well defined, the stress and strain during the load application are known also in the critical locations where the crack starts. These points belong to the external surface of the specimens in the pure torsion and in mixed condition (torsion plus tension/compression) tests, while they are located in the centre of the specimen for tensile tests. It is important to state again that the calibration is not based on a single value but on the ratio between the strain increment and the failure strain at the same stress triaxiality. Moreover each test involves a range of stress triaxiality, thus the calibration process consists in the finding of the equation parameter which causes the failure of the critical elements at the desired step. After the right limit curve has been fitted, it can be implemented like tabular ductile damage criterion into the ABAQUS 6.8 software. The numerical analyses are then re-performed to compare the numerical curves with the experimental ones and finally the behaviour in terms of physical quantities are fully evaluated, Fig. 9. At this point two different zones are considered in order to calibrate the fracture locus:

1. For the tensile tests (high stress triaxility, >0.4) a hyperbolic equation for the fracture locus is supposed, thus only one unknown quantity must be calibrated. Referring to Fig. 4, this zone is the one in which ductile failure is shown and it considers high values of η.
2. For the low stress triaxiality ($0 \le \eta \le 0.4$) tests it is hard to find a simple fracture locus curve that covers all the different situations. Referring to Fig. 4, in this zone a combination of shear and ductile failure appears. The approach followed in this work is identical to Bao and Wierzbicki [1, 33]. This means that each numerical test covers a point of the fracture locus. The points are obtained using the average stress triaxiality and the equivalent plastic strain at the critical rotation. The average stress triaxiality η_{av} has been calculated as:

$$\eta_{av} = \frac{1}{\varepsilon_f} \int_0^{\varepsilon_f} \eta d\varepsilon_{pl} \tag{6}$$

Once all the points are collected it is possible to build the fracture locus. In the present chapter the points are linked by a straight line and finally the fracture locus is implemented into ABAQUS and the analysis is restarted. Although the choice of linking all the points by lines might seem a simplified method, it provides very good results indeed.

As far as negative stress triaxiality is concerned, the upsetting tests show no fracture. This is compliant with the fracture locus curve with a vertical asymptotic value for a stress triaxiality of $-1/3$. Moreover for the negative range of triaxialities the result of the multiaxial test with torsion-compression is also available. Unfortunately this test doesn't result in failure of the specimen even after a large rotation (270°). Thus the fracture locus has been built with a hyperbolic equation under the following constraints: (1) to be asymptotic for a stress triaxiality of $-1/3$ and (2) to have the same value obtained from the pure torsion test (zero stress triaxiality). In addiction the hyperbolic equation has to be higher than the experimental point obtained from the torsion-compression test. In Table 8 the results of the fracture locus are summarized. The curve is divided into different stress triaxiality and for each range the equation of the fracture locus is provided.

A general description of the entire work that has been carried out can be seen in Fig. 10. This figure reports the evolution of the plastic strain ε_{pl} (vs. triaxiality), obtained from each numerical simulation of the experimental tests. The final value of ε_{pl} is the plastic strain at failure ε_f. However, as previously underlined, the fracture locus curve is not simply the curve that fits at best these points. During tests in fact the stress triaxiality doesn't remain constant but changes and the failure is obtained through the integration of the ratio of the strain increment with the failure strain at the same stress triaxiality, Eq. (3), during the entire test. Therefore in order to reduce the complexity of the procedure it is necessary to define the average stress triaxiality Eq. (6). The final data obtained from each test thus is a point (η_{av}, ε_f) which is necessary in order to build a complete fracture locus. In Fig. 11 the black cross, the points (η_{av}, ε_f) found with virtual testing procedure, are presented with the final fracture locus curve for aluminium Al 6061-T6 alloys. Figure 11 shows a comparison between this new fracture locus and a JC fracture locus reported in the literature [22].

8 Validation of the Fracture Model in High Velocity Impact

The fracture locus and the constitutive law calibrated in the previous paragraphs are then analysed in other scenarios in order to verify the transferability of the results. The main aim of the current research is to find a reliable flow stress law and a reliable fracture locus for high velocity impacts and, hence, the validation considers principally this field. However, as previous remarked, the parameters taking the strain rate behaviour into account have been obtained from literature [22, 34, 35].

8.1 Normal Impact of a Rigid Cylinder on a Plate

The final purpose of the calibration is to simulate impact problems. Hence, the first way to validate the material parameters is to reproduce simple impact problems with a rigid impactor. The experimental data are extracted from Li and Goldsmith's work [36]. The authors carried out a series of experimental impacts with a steel blunt cylinder against an Al 6061-T6 aluminium plate. The main result of Li and Goldsmith is the description of the residual velocity of the cylinders after the perforation of the plates. Thus a model has been built with the commercial finite elements software ABAQUS. Using the symmetry condition, only a quarter of the plate is required to be built. The impacting cylinder is considered to be rigid. The plate has been meshed with

Table 8 Definition of the new fracture locus calibrated for Al 6061-T6

Stress triaxiality η	$\eta_{av} < 0$		$0 < \eta_{av} < 0.0223$		$0.0223 < \eta_{av} < 0.0626$		$0.062 < \eta_{av} < 0.37$		$\eta_{av} > 0.37$	
Equation	$\varepsilon_f = \frac{A}{\eta+1/3} - 3A + \varepsilon_{ftp}$		$\varepsilon_f = m \cdot \eta + q$		$\varepsilon_f = m \cdot \eta + q$		$\varepsilon_f = m \cdot \eta + q$		$\varepsilon_f = \frac{A}{\eta}$	
Coefficients	**A**	0.428	**m**	20.85	**m**	−5.43	**m**	−0.848	**A**	0.17
	ε_{ftp}	0.474	**q**	0.474	**q**	1.060	**q**	0.774		

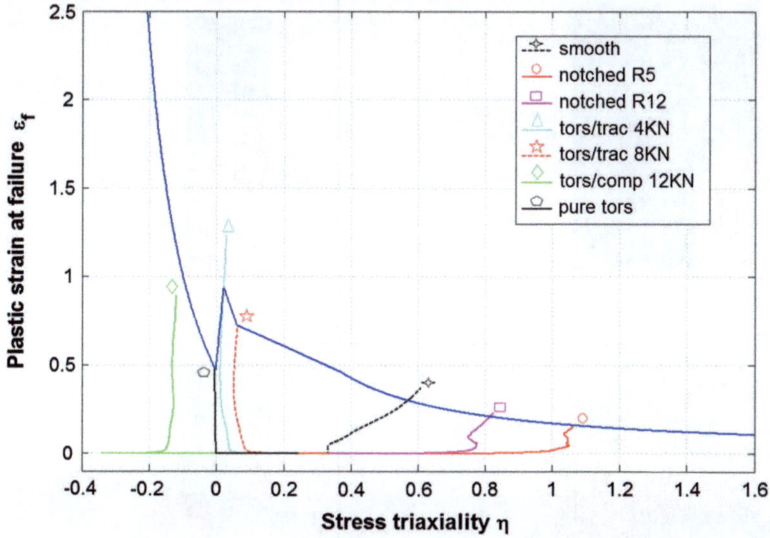

Fig. 10 New calibrated fracture locus with the evolution of plastic strain versus stress triaxiality for each test until failure

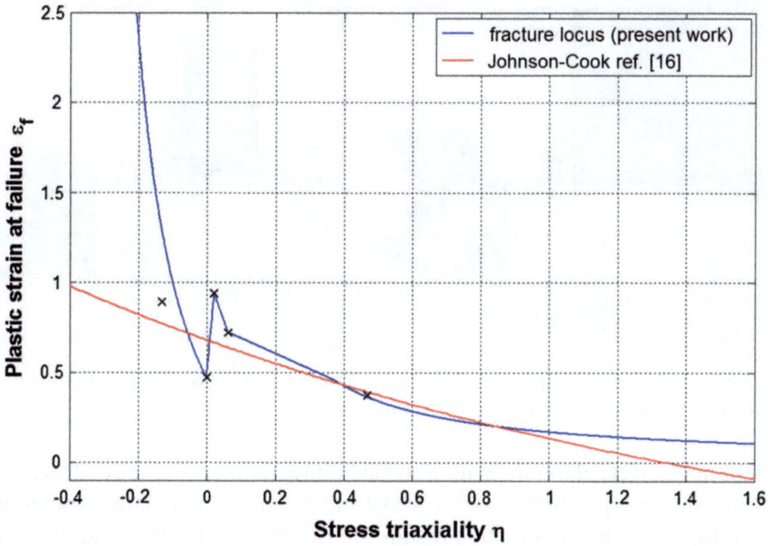

Fig. 11 Comparison between the new calibrated fracture locus for Al 6061-T6 and a literature Johnson–Cook criterion

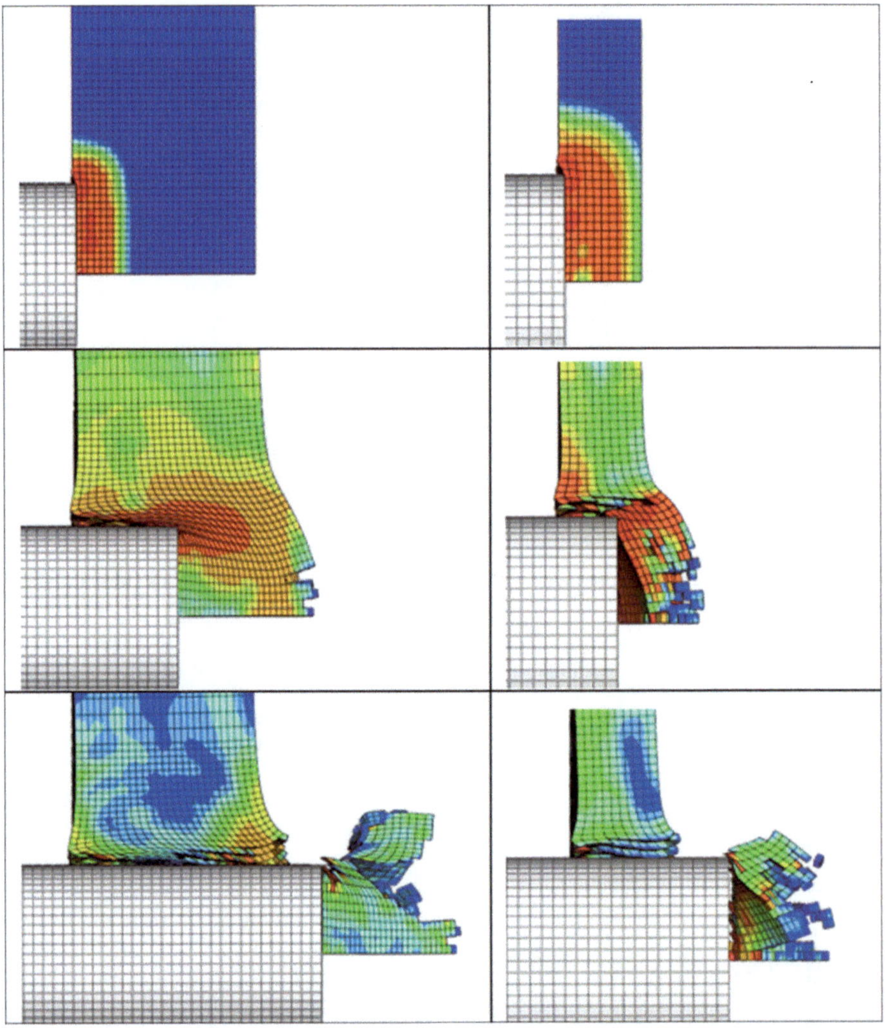

Fig. 12 Impact of a rigid cylinder with a flat nose against an Al 6061-T6 aluminium alloy plate with different thickness: 12.8 mm (*left*), 4.8 mm (*right*)

solid eight nodes type elements with a reduced integration scheme (C3D8R) with an average dimension equal to $0.5 \times 0.5 \times 0.47$ mm. The results of the simulations can be seen in Fig. 12. The quantitative results are reported in Table 9 and they stand in good agreement with the experimental data. The error between the numerical and experimental residual velocity is always below 10 %, with an average error among all the tests of 3.1 %. Another important result is the presence of limited distortion of the elements involved during the impact simulation. Comparing the results of these analyses with the equivalent material calibration found in the literature [22], a large number of distorted elements are present in the latter analyses.

Table 9 Comparisons between numerical and experimental results for the impact of a rigid cylinder with a flat nose against an Al 6061-T6 aluminium alloy plate

Plate thickness (mm)	Impact velocity (m/s)	Experimental [36] Residual velocity (m/s)	Numerical Residual velocity (m/s)	Error (%)
4.8	409	392	383	−2.3
4.8	525	502	497	−1.0
4.8	565	543	535	−1.5
12.8	402	323	294	−9.0
12.8	565	478	469	−1.9

8.2 Ballistic Impact of a 7.62 NATO Bullet on a Transmission Shaft

As discussed in the introduction, such a detailed calibration of the material behaviour is aimed to numerically simulate mechanical components under extreme load. In particular the authors are at present focusing on the assessment of a helicopter transmission shaft when subjected to ballistic impact. A numerical model, able to reproduce correctly the consequent damage caused by an impact, could be useful as base layer for a subsequent damage tolerant analysis, which in turn allows the evaluation of the residual integrity of the shaft. Due to the complexity of the problem an experimental campaign has been carried out by the authors [37] and finally the experimental data have been compared with the results from the numerical simulations. The shaft has been modelled with a refined mesh size near the impact zone: four elements in the thickness with dimensions of $0.75 \times 0.5 \times 0.625$ mm, C3D8R type. The projectile is a 7.62 NATO calibre which has been assumed to be rigid. The impact condition considered in the tests is the tangential 45° oriented impact (with respect to the longitudinal axis) which, according to Colombo and Giglio [15], is the most dangerous. It causes damage in the shape of one single large hole. The numerical calibration of the material previously illustrated provides reliable results in terms of residual velocity with errors lower than 10 % and an encouraging comparison of the shape of the damage [38]. Moreover, Fig. 13 shows a qualitative comparison between two analyses obtained with a fracture locus JC type from the literature [22] and the new calibrated fracture locus. Excessive element distortion is present in the JC analyses which are therefore unable to fully represent realistic failure for all the different stress conditions. The analyses using the calibrated fracture locus show, on the contrary, a much more realistic damage. In Fig. 14 a picture superimposing a 3D scan of the damage (red line) with the FE results is shown. The 3D scan shows a more extensive damage in the experimental impacts that is likely to be caused by the deformation of the bullet, a parameter which has not been taken into account in

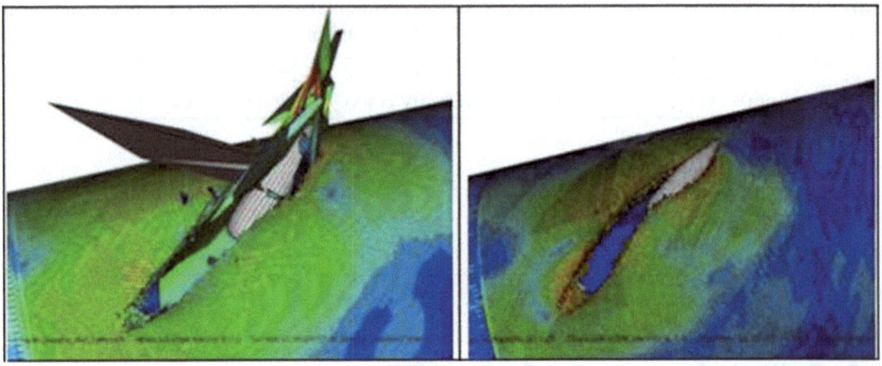

Fig. 13 Excessive distorted elements with the literature fracture locus (*left*), correct analysis with the new calibrated fracture locus (*right*)

Fig. 14 Acquired boundary curve (*red line*) projected on the damage simulation model

this analysis (the bullet has been reproduced as a rigid body). However, the longitudinal extension of the failure and the entry area of the damage (lower right in the Fig. 14) are well reproduced. At this stage the importance of the calibration of a fracture locus over a large range of stress triaxiality needs to be underlined. Stress triaxiality on a component subjected to a ballistic impact can in fact vary from negative to positive value. Figure 15 shows the stress triaxiality referring to an element near the impacted zone. This element suffers a high oscillating triaxiality up to the time when it is eliminated from the analysis due to failure. From the instant of failure, stiffness is set to zero thus also triaxiality is set to zero.

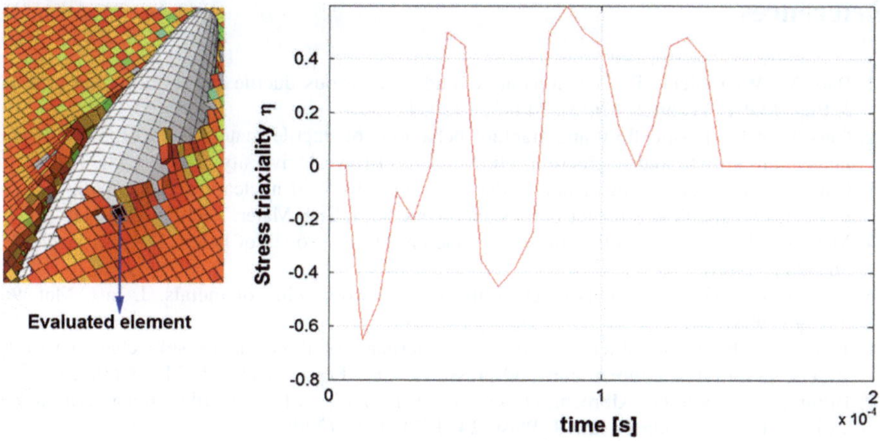

Fig. 15 Stress triaxiality of an element near the impacted zone

9 Conclusive Remarks

In the present chapter a complete material model for an Al 6061-T6 aluminium alloy has been built. The calibration accuracy is evaluated comparing numerical impact models that stand in very good agreement with experimental tests. Reliable results in an extreme condition phenomenon such as a ballistic impact strongly depend on the use of a proper calibrated material law. This law needs to be not only consistent for a limited but also for a wide range including hardening and fracture locus. In this scenario the effect of triaxiality on fracture is a crucial parameter. For such analyses, especially as prediction of the range of stress triaxiality involved during a ballistic impact are almost impossible to make. Thus, in order to achieve a comprehensive calibration of the material, a multiaxial tests procedure has been used by the authors. Thanks to this procedure a better understanding of the behaviour of the material, especially with regards to the shape of the fracture locus in the low stress triaxiality range ($0 \leq \eta \leq 0.4$), has been obtained. The shape in this zone is under debate in the literature due to a "discontinuity" near the zero value (pure shear). This particular discontinuous behaviour has been further confirmed in this work. Besides, many efforts have been made to develop numerical virtual simulations of all the experimental tests. The validation of the calibrated damage model in complex impacts simulations shows the large potential of the developed methodology.

References

1. Bao, Y., Wierzbicki, T.: A comparative study on various ductile crack formation criteria. J. Eng. Mater. Technol. **126**(3), 314–324 (2004)
2. Gurson, A.L.: Plastic flow and fracture behaviour of ductile materials incorporating void nucleation, growth and interaction. Ph.D. Thesis, Brown University (1975)
3. Gurson, A.L.: Continuum theory of ductile rupture by void nucleation and growth: part 1-yield criteria and flow rules for porous ductile media. J. Eng. Mater. Technol. **99**, 2–15 (1977)
4. McClintock, F.A.: A criterion of ductile fracture by the growth of holes. J. Appl. Mech. **35**, 363–371 (1968)
5. Cockcroft, M.G., Latham, D.J.: Ductility and the workability of metals. J. Inst. Met. **96**, 33–39 (1968)
6. Johnson, G.R., Cook, W.H.: Fracture characteristics of three metals subjected to various strains, strain rates, temperatures and pressures. Eng. Fract. Mech. **21**, 31–48 (1985)
7. Brunig, M., Chyra, O., Albrecht, D., Driemeier, L., Alves, M.: A ductile damage criterion at various stress triaxialities. Int. J. Plast. **24**, 1731–1755 (2008)
8. Coppola, T., Cortese, L., Folgarait, P.: The effect of stress invariants on ductile fracture limit in steels. Eng. Fract. Mech. **76**, 1288–1302 (2009)
9. Bonora, N.: A non-linear CDM model for ductile failure. Eng. Fract. Mech. **58**, 11–28 (1997)
10. Lemaitre, J.: A continuous damage mechanics model for ductile fracture. J. Eng. Mater. Technol. **107**, 83–89 (1985)
11. Bonora, N., Ruggero, A., Esposito, L., Gentile, D.: CDM modeling of ductile failure in ferritic steels: assessment of the geometry transferability of model parameters. Int. J. Plast. **22**, 2015–2047 (2006)
12. Bonora, N., Gentile, D., Pirondi, A., Newaz, G.: Ductile damage evolution under triaxial state of stress: theory and experiments. Int. J. Plast. **21**, 981–1007 (2005)
13. Chiantoni, G., Bonora, N., Ruggero, A.: Experimental study of the effect of triaxiality ratio on the formability limit diagram and ductile damage evolution in steel and high purity copper. Int. J. Mater. Form. **3**(1), 171–174 (2010)
14. www.matweb.com
15. Colombo, D., Giglio, M.: Numerical analysis of thin-walled shaft perforation by projectile, pp. 1264–1280. Computers & Structures, Elsevier Applied Science (2007)
16. Follansbee, P.S., Kocks, U.F.: A constitutive description of the deformation of copper based on the use of the mechanical threshold stress as an internal state variable. Acta Metall. Mater. **36**(1), 81–93 (1988)
17. Goto, D.M., Bingert, J.F., Chen, S.R., Gray III, G.T., Garrett Jr, R.K.: The mechanical threshold stress constitutive strength model description of HY-100. Metall. Mater. Trans. A **31a**, 1985–1996
18. Campagne, L., Daridon, L., Ahzi, S.: A physically based model for dynamic failure in ductile metals. Mech. Mater. **37**, 869–886 (2005)
19. Clausen, A.H., Börvik, T., Hopperstad, O.S., Benallal, A.: Flow and fracture characteristics of aluminium alloy AA5083-H116 as function of strain rate, temperature and triaxiality. Mater. Sci. Eng. A **364**, 260–272 (2004)
20. Børvik, T., Hopperstad, O.S., Berstad, T., Langseth, M.: A computational model of viscoplasticity and ductile damage for impact and penetration. Eur. J. Mech. A-Solid **20**, 685–712 (2001)
21. Johnson, G.R., Holmquist, T.J.: Test data and computational strength and fracture model constants for 23 materials subjected to large strain, high-strain rates, and high temperatures. Technical Report LA-11463-MS, Los Alamos National Laboratory (1989)
22. Lesuer, D.R., Kay, G.J., LeBlanc, M.M.: Modeling large-strain, high rate deformation in metals. UCRL-JC-134118, Lawrence Livermore National Laboratory (2001)
23. Zukas, J.A.: High Velocity Impact Dynamics. Wiley, London (1990)
24. Meyers, M.A.: Dynamic Behavior of Materials. Wiley, New York (1994)

25. Bao, Y., Wierzbicki, T.: Bridgman revisited: on the history effects on ductile fracture. J. Mech. Phys. Solids (2004)
26. Clausing, D.P.: Effect of plastic strain state on ductility and toughness. Int. J. Fract.Mech. **6**, 71–85 (1970)
27. McClintock, F.A.: Plasticity aspects of fracture. In: Liebowitz, H. (Ed.) Fracture, Vol. 3, pp 47-225. Academic Press, New York (1971)
28. Mirone, G.: Role of stress triaxiality in elastoplastic characterization and ductile failure prediction. Eng. Fract. Mech. **74**, 1203–1221 (2007)
29. Driemeier, L., Brunig, M., Micheli, G., Alves, M.: Experiments on stress-triaxiality dependence of material behavior of aluminum alloys. Mech. Mater. **42**, 207–217 (2010)
30. Li, H., Fu, M.W., Lu, J., Yang, H.: Ductile fracture: experiments and computations. Int. J. Plast. **27**, 147–180 (2011)
31. Hopperstad, O.S., Börvik, T., Langseth, M., Labibes, K., Albertini, C.: On the influence of stress triaxiality and strain rate on the behaviour of a structural steel. Part I: experiments. Eur. J. Mech. A-Solid **22**, 1–13 (2003)
32. Börvik, T., Hopperstad, O.S., Berstad, T., Langseth, M.: Numerical simulation of plugging failure in ballistic penetration. Int. J. Solids Struct. **38**(34–35), 6241–6264 (2001)
33. Bao, Y., Wierzbicki, T.: On the cut-off value of negative stress triaxiality for fracture. Eng. Fract. Mech. **72**, 1049–1069 (2005)
34. Dabboussi, W., Nemes, J.A.: Modeling of ductile fracture using the dynamic punch test. Int. J. Mech. Sci. **47**, 1282–1299 (2005)
35. Zukas, J.A., Nicholas, T., Swift, H.F.: Impact Dynamics. Wiley, New York (1982)
36. Li, K., Goldsmith, W.: Impact on aluminium plates by tumbling projectiles: experimental study. Int. J. Impact Eng. **18**(1), 23–43 (1996)
37. Giudici, L., Manes, A., Giglio, M.: Ballistic impact on a tail rotor transmission shaft for helicopter. In: International Conference Ballistic 2010, Beijing, 17–21 May 2010 (China)
38. Manes, A., Magrassi, G., Giglio, M., Bordegoni, M.: Reverse engineering of experimental tests results of ballistic impact for the validation of finite element simulations. In: ASME 2010 International Design Engineering Technical Conferences and Computers and Information in Engineering Conference IDETC/CIE (2010)

An Improved Finite Element Model for Numerical Simulation of Phase Changes of Iron Under Extreme Conditions

Ladislav Écsi, Pavel Élesztős and Kinga Balázsová

Abstract In this chapter an improved finite element model for numerical simulation of phase changes of iron is presented, which is capable of simulating iron construction behaviour under extreme conditions that include large strains/large deformations at high strain rates and temperatures. The model is based on an improved variational formulation of the conservation of energy with convective heat transfer. It employs the updated Lagrangian formulation and uses the extended NoIHKH material model for cyclic plasticity of metals. It also uses the Kelvin–Voigt model for internal damping, the Jaumann objective rate in the Cauchy's stress update calculation and simplified rate forms of the Mehl-Avrami and Koisten-Marburger equations for ferrite, pearlite, bainite, martensite and the retaining austenite phase calculation. A numerical experiment using a cooled bar in cyclic bending is presented and briefly discussed.

Keywords FEM · Thermal-structural analysis · Phase changes · Cyclic plasticity of metals · Updated Lagrangian formulation

1 Introduction

In many applications, especially in the space industry (e.g. jet engines and rockets) as well as electronic industry (e.g. superheat resistant materials for thermal protection, vehicle and personal body armour, electromagnetic sensors), structures or part of structures are exposed to high temperatures, high temperature gradients and/or cyclic temperature changes or impact loading [1]. Although conventional metallic materials cannot operate above certain temperatures for long as they would change their material properties due to changes in their microstructure

L. Écsi (✉) · P. Élesztős · K. Balázsová
Faculty of Mechanical Engineering, Slovak University of Technology in Bratislava, Námestie Slobody 17, 812 31 Bratislava, Slovakia
e-mail: ladislav.ecsi@stuba.sk

N. Bonora and E. Brown (eds.), *Numerical Modeling of Materials Under Extreme Conditions*, Advanced Structured Materials 35, DOI: 10.1007/978-3-642-54258-9_8,

caused by creep, plastic deformations, phase transformations or grain coarsening etc., in order to ensure their safe operation and reliability, it is essential to understand their material behaviour at elevated temperatures, mainly if their deformation takes place at high strain rates. From the thermodynamics point of view each body that experiences a loading described in the above, represents a non-equilibrium thermodynamic system with ongoing internal processes that interact with each other and might generate or absorb heat.

Numerical simulations of the behaviour of a solid deformable body outside thermodynamic equilibrium thus represent a challenging task in computational mechanics that needs careful consideration.

While the thermostatics that describes the transition from a state of equilibrium to another state of equilibrium was developed in an elegant mathematical form by Clausius, Gibbs, Duhem and Carathéodory in agreement with experiments, unfortunately the same cannot be said about the thermodynamics outside equilibrium [2]. Although thermodynamical aspects of mathematical modelling of continuous media are now well-established [3–6] and numerous mathematical models and techniques have been developed [7–20], which have resulted from extensive studies and observations of various phenomena in a deforming material, contemporary finite element analyses of coupled thermal structural problems use heat equation exclusively [2, 21–23] to describe the conservation of energy. The equation usually contains a mechanical coupling term that defines intrinsic dissipation. Such a formulation can model deformable body behaviour in all cases where temperatures are directly prescribed at some material points of the body, as it is the case with prescribed temperatures on the boundary.

If however convective or radiative heat transfer is applied, the heat equation cannot be solved alone using a conventional solution even if there is no mechanical coupling term in it, as the temperature field and the displacement field of the body is coupled via the Cauchy's stress theorem on the boundary. Until now the issue has been disregarded in weak formulations which eventually lead to either a one-way or two-way coupled sequential finite element analyses, depending on if small strain/finite strain theory was used. In order to address the coupling via BC, the convective and the radiative heat transfer need a somewhat different treatment from the weak solution point of view [24] which we will discuss in detail later.

Furthermore, certain internal dissipation processes, such as internal/material damping are still neglected in contemporary finite element formulations, which might prevent the analysis to converge to a more realistic solution. This is especially true if the body undergoes large deformations at high strain rates, which above the recrystallization temperature is accompanied by phase changes too [25].

Over the past decades great effort has been spent on developing suitable material models to describe deformable solid body behaviour under extreme conditions. They either use rate independent theories of plasticity [26–31] combined with damage [32], for non-metallic materials, metal-ceramic composites and functionally graded materials, as it is described in [1] or rate dependent theories with and without damage evolution [26–30], particularly the theory of viscoplasticity that employs temperature, strain and strain rate dependent evolution equations to formulate viscoplastic strain rate tensors [33–35], or/and flow stresses [36–44] for materials at high strain rates. The greatest disadvantages of the above models are that they do not dissipate energy in elastic loadings. Although there are quite a few more complex material models that solve the problem by combining viscoelastic and viscoplastic theories, they have been proposed for slow strain rates [45, 46], to model creep [47–50], shape memory alloys [51] that might be combined with transformation induced plasticity at room temperature [52, 53] and phase transformations above the material recrystallization temperature [54, 55]. The models however are too complex from the material parameter identification point of view to be used for practical analyses [56–58].

The objective of the work is to develop a simple finite element model for fully coupled thermal-structural finite element analyses with convective and radiative heat transfer and phase transformations that within the framework of large strain/large deformation formulation considers the strong coupling between the temperature field and the displacement field on the boundary of a solid deformable body, that is outside thermodynamic equilibrium, and takes into account the internal damping of the material. The presented work by no means is considered to be complete and exhaustive as there are too many simplifications in it, but according to the authors belief it represents a possible way of improving the accuracy of contemporary finite element analyses of the above problem, as it uses improved weak formulation and considers internal damping of the material, thus it also enables simulating iron construction behaviour under extreme conditions.

2 Theory Background

Our research focuses on the study of the behaviour of a solid deformable body by means of fully coupled thermal-structural analysis using the finite element method. We assume that the modelled material is initially homogenous and isotropic and that the body loading is purely mechanical. The only heat exchange between the body and its surroundings is realized via convective heat transfer. Our aim is to keep the applied theory as simple as possible and rely on well established and approved material models with as few as possible easily identifiable parameters, which are key aspects of the mathematical model acceptability for practical engineering calculation purposes. In the centre of our attention are metals, which as the most widely used construction materials, have extensively been studied in the past decades [22, 28–32] which eventually resulted in the development of the well known classes of material models [59–61].

2.1 Weak Form for the Updated Lagrangian Formulation

In the presented work we assume that a solid deformable body, idealized as non-polar continuum, undergoes large deformations. In the used energy conservation equation derivation we kept strictly to the basic principles of thermodynamics. We neither strived for completeness nor tried to include any dissipative process into the analysis, which might have taken place in the material, but its mathematical formulation has not yet been clarified or verified experimentally. Considering the updated Lagrangian formulation for finite strains, the governing equations (1)–(5), i.e. the balance of mass, the balance of linear momentum, the balance of energy and the applied boundary conditions (BC) can be expressed with the following formulas:

$$\rho_0 = \rho J, \tag{1}$$

$$\mathbf{b} + \nabla \cdot \boldsymbol{\sigma} - \rho \dot{\mathbf{v}} = \mathbf{0}, \quad \boldsymbol{\sigma} = \boldsymbol{\sigma}^{\mathrm{T}}, \tag{2}$$

$$\mathbf{v} \cdot (\rho \dot{\mathbf{v}} - \mathbf{b} - \nabla \cdot \boldsymbol{\sigma}) + \dot{e} - (\boldsymbol{\sigma} : \mathbf{d} - \nabla \cdot \mathbf{q} + r) = 0, \tag{3}$$

BC:

$$-\mathbf{q} \cdot \mathbf{n} = q_n = h(T_{BULK} - T_S), \tag{4}$$

$$\boldsymbol{\sigma} \cdot \mathbf{n} = \mathbf{t}. \tag{5}$$

Here $\boldsymbol{\sigma}, \mathbf{d}, \mathbf{q}, \mathbf{b}, \mathbf{v}, \mathbf{t}, \mathbf{n}, e, r, q_n, \rho, \rho_0, J, h, T_S, T_{BULK}$ denote the Cauchy stress tensor, the strain rate tensor, the heat flux vector, the body force vector, the velocity vector, the surface traction vector, the unit outward surface normal vector, the internal energy per unit volume, the heat generation rate per unit volume, the normal heat flux, the material density in current and initial configuration, the Jacobian or the determinant of the deformation gradient, the heat transfer coefficient, the surface temperature and the bulk temperature respectively. Referring to the formulation of the balance of energy (3), it is necessary to mention here the difference between conventional and weak solution of the governing system of partial differential equations (PDE) (2), (3) [62, 63]. Equation (3) is the sum of the conservation of mechanical energy and a heat equation, and since the conservation of mechanical energy is the scalar product of the balance of linear momentum (2) and the velocity at a material point, the balance of energy can be replaced with the heat equation, if conventional solution is used [24]. Moreover, regardless of the final formulation of the heat equation, if certain types of boundary conditions, such as convective heat transfer (4) or radiative heat transfer [64, 65] are applied, Eq. (5) that defines a two-way coupling between the displacement field and the temperature field on the boundary, needs to be taken into account in the problem weak formulation [24]. During the weak solution Eqs. (4) and (5) are not solved directly, only their left-hand sides are replaced with their right-hand sides in the energy functional appropriate term integration. Then the only way to take into account the coupling via (5) is to include the conservation of mechanical energy into the weak form of the balance of energy (7) [24, 66–69], which changes the finite element formulation into a constrained minimization problem [70, 71]. After substituting the final formula of the heat equation into Eq. (3) and applying the Galerkin's method to the governing PDE, the weak form for the updated Lagrangian formulation can be derived [24, 67]:

$$\int_{\Omega} \rho\dot{\mathbf{v}} \cdot \delta\mathbf{v} dv + \int_{\Omega} \boldsymbol{\sigma} : \delta\mathbf{d} dv = \int_{\Omega} \mathbf{b} \cdot \delta\mathbf{v} dv + \int_{\partial\Omega} \mathbf{t} \cdot \delta\mathbf{v} ds, \tag{6}$$

$$\begin{aligned} &\int_{\Omega} \delta T \rho\dot{\mathbf{v}} \cdot \mathbf{v} dv + \int_{\Omega} \delta T(\boldsymbol{\sigma} : \mathbf{d}) dv + \int_{\Omega} (\nabla\delta T) \cdot (\boldsymbol{\sigma} \cdot \mathbf{v}) dv + \int_{\Omega} \delta T \rho c \dot{T} dv \\ &- \int_{\Omega} (\nabla\delta T) \cdot \mathbf{q} dv = \int_{\Omega} \delta T \mathbf{b} \cdot \mathbf{v} dv + \int_{\partial\Omega} \delta T \mathbf{t} \cdot \mathbf{v} ds + \int_{\partial\Omega} \delta T q_n ds + \int_{\Omega} \delta T r dv, \end{aligned} \tag{7}$$

where $\delta\mathbf{d}, \delta\mathbf{v}, \delta T$ stand for the virtual strain rate tensor, virtual velocity vector and virtual temperature respectively. In Eq. (7) we formally used the simplest formula for the heat equation and considered the thermal conduction law of Fourier type $\mathbf{q} = -\mathbf{K} \cdot (\nabla T)$, where $\mathbf{K}$ denotes the conductivity tensor and c is the specific heat capacity. In spite of the simplest formula we can say that the heat equation is complete and applicable for both, elastic and plastic analysis, if we reinterpret the meaning of the heat generation rate per unit volume r in it in the sense that any heat generated during the deformation of the body can be viewed as a heat generation rate; then any intrinsic dissipation process can also be considered a heat generation rate [2].

2.2 Equivalence Between the Strong and Weak Equilibrium Statements Damping

When the stress field $\boldsymbol{\sigma}$, heat flux field $\mathbf{q}$ and the velocity field $\mathbf{v}$ are sufficiently smooth, the weak equilibrium statements Eqs. (6) and (7) are equivalent with the strong equilibrium statements Eqs. (2)–(5). To show this let us follow the procedure described in [84]. Assuming that $\boldsymbol{\sigma}, \mathbf{q}, \mathbf{v}$ are sufficiently regular, the following identities apply

$$\boldsymbol{\sigma} : \delta\mathbf{d} = \nabla \cdot (\boldsymbol{\sigma} \cdot \delta\mathbf{v}) - (\nabla \cdot \boldsymbol{\sigma}) \cdot \delta\mathbf{v}, \tag{8}$$

$$\nabla\delta T \cdot (\boldsymbol{\sigma} \cdot \mathbf{v}) = \nabla \cdot [\delta T(\boldsymbol{\sigma} \cdot \mathbf{v})] - \delta T \nabla \cdot (\boldsymbol{\sigma} \cdot \mathbf{v}), \tag{9}$$

$$\nabla \delta T \cdot \mathbf{q} = \nabla \cdot (\delta T \mathbf{q}) - \delta T \nabla \cdot \mathbf{q}, \tag{10}$$

$$-\nabla \cdot (\boldsymbol{\sigma} \cdot \mathbf{v}) = -\boldsymbol{\sigma} : \mathbf{d} - (\nabla \cdot \boldsymbol{\sigma}) \cdot \mathbf{v}. \tag{11}$$

After substituting (8)–(11) into (6), (7) and considering that the internal energy time derivative for this particular case is $\dot{e} = \boldsymbol{\sigma} : \mathbf{d} + \rho c \dot{T}$ [24], the weak equilibrium statements can be manipulated into the following forms:

$$\int_{\Omega} \delta \mathbf{v} \cdot (\rho \dot{\mathbf{v}} - \nabla \cdot \boldsymbol{\sigma} - \mathbf{b}) dv + \int_{\partial\Omega} \delta \mathbf{v} \cdot (\boldsymbol{\sigma} \cdot \mathbf{n} - \mathbf{t}) ds = 0, \tag{12}$$

$$\begin{aligned} &\int_{\Omega} \delta T [\mathbf{v} \cdot (\rho \dot{\mathbf{v}} - \nabla \cdot \boldsymbol{\sigma} - \mathbf{b}) + (\dot{e} - \boldsymbol{\sigma} : \mathbf{d} + \nabla \cdot \mathbf{q} - r)] dv \\ &\quad + \int_{\partial\Omega} \delta T [\mathbf{v} \cdot (\boldsymbol{\sigma} \cdot \mathbf{n} - \mathbf{t})] ds + \int_{\partial\Omega} \delta T (-\mathbf{q} \cdot \mathbf{n} - q_n) ds = 0. \end{aligned} \tag{13}$$

Finally, since Eqs. (12) and (13) hold for all virtual velocity fields $\delta \mathbf{v}$ and all virtual temperature fields δT, then it follows from the fundamental theorem of variational calculus [72–74] that each bracketed term of the above equations must vanish pointwise within the respective domain, i.e. we recover the strong equilibrium equations (2)–(5). As Eqs. (6), (7) and (12), (13) are equivalent, from the later two the following can be asserted:

1. From the physical point of view the weak form of balance of energy (7) is more complete as it contains all mechanical energy contributions, i.e. the balance of total energy of the body is the sum of the balance of mechanical energy and the balance of heat energy.
2. From the mathematical point of view the coupling via BC (5), or via the application of the Cauchy's stress theorem on the boundary is enforced in both weak forms (6) and (7).
3. In limiting state, when the velocity $\mathbf{v}$ approaches zero and taking into account the internal energy time derivative definition above, the sequential finite element solution is recovered.

Moreover, since the conservation of mechanical energy can also be viewed as a constraint equation for temperature field determination in (7), any direct temperature imposition will result in an over-constrained system of algebraic finite element equations with no solution for the sought nodal displacements and temperatures.

2.3 Constitutive Equation with Internal Damping

Material or internal damping plays an important role in deformable body behaviour if the loading of the body is dynamic or the body undergoes small/large deformations at high strain rates. Although technical literature differentiates between material or structural viscous damping, nonmaterial viscous damping, Coulomb or dry damping and other nonmaterial damping models [75], the material damping is merely an issue of rate dependent models of viscoelasticity or viscoplasticity which have been developed to simulate materials with memory, creep or relaxation using convolution integrals [28, 76, 77]. The Rayleigh damping, or in other words the proportional damping, is the most widely used damping model in contemporary dynamic analyses [78–81], however it has to be carefully used in continuum based models, as the damping model does not differentiate between rigid body motion and deformation which might lead to spurious numerical behaviour and inaccurate results. This applies mainly in such cases where high speed rigid body motion takes place during the deformation of the body. Considering the co-rotational formulation [82] adjustment of the Kelvin-Voigt model [24] for finite strain elastoplasticity [83] results in the co-rotational Cauchy's stress tensor $\hat{\sigma} = \mathbf{R}^T \cdot \boldsymbol{\sigma} \cdot \mathbf{R}$, Eq. (14) to be split into two objective stress measures, represented by an incrementally changing elastic part or co-rotational elastic Cauchy's stress tensor $\hat{\sigma}^{el}$, Eq. (15), and an instantaneous damping part or co-rotational viscous Cauchy's stress tensor $\hat{\sigma}^{damp}$, Eq. (18), where the co-rotational elastic Cauchy's stress increment $\Delta\boldsymbol{\theta}^{el}$ is defined either with Eq. (16) in elastic loading/unloading or with Eq. (17) in plastic loading.

$$^{n+1}\hat{\sigma} = {}^{n+1}\hat{\sigma}^{el} + {}^{n+1}\hat{\sigma}^{damp}, \tag{14}$$

where

$$^{n+1}\hat{\sigma}^{el} = \Delta\hat{\sigma}^{el} + {}^{n}\hat{\sigma}^{el}, \tag{15}$$

$$\Delta\hat{\sigma}^{el} = \Delta t^{n+\frac{1}{2}}\hat{\mathbb{C}}^{\nabla J} : \left({}^{n+\frac{1}{2}}\hat{\mathbf{d}} - {}^{n+\frac{1}{2}}\hat{\mathbf{d}}th\right), \text{ in elastic loading/unloading} \tag{16}$$

$$\Delta\hat{\sigma}^{el} = \Delta t^{n+\frac{1}{2}}\hat{\mathbb{C}}^{\nabla J} : \left({}^{n+\frac{1}{2}}\hat{\mathbf{d}} - {}^{n+\frac{1}{2}}\hat{\mathbf{d}}^{th} - {}^{n+\frac{1}{2}}\hat{\mathbf{d}}^{pl}\right), \text{ in plastic loading} \tag{17}$$

$$^{n+1}\hat{\sigma}^{damp} = {}^{n+1}\hat{\mathbb{C}}^{\nabla Jdamp} : {}^{n+1}\hat{\mathbf{d}}, \tag{18}$$

$$^{0}\mathbb{C}^{\nabla J} = 2G + \lambda(\mathbf{1} \otimes \mathbf{1}),\ G = \frac{E}{2(1+v)},\ \lambda = \frac{vE}{(1+v)(1-2v)}, \tag{19}$$

$$^{n+\frac{1}{2}}\hat{\mathbb{C}}^{\nabla J}_{abcd} = \frac{1}{^{n+\frac{1}{2}}J}\,{}^{n+\frac{1}{2}}R_{ma}\,{}^{n+\frac{1}{2}}R_{nb}\,{}^{n+\frac{1}{2}}R_{oc}\,{}^{n+\frac{1}{2}}R_{pd}\,{}^{n+\frac{1}{2}}F_{mi}\,{}^{n+\frac{1}{2}}F_{nj}\,{}^{n+\frac{1}{2}}F_{ok}\,{}^{n+\frac{1}{2}}F_{pl}\,{}^{0}\mathbb{C}^{\nabla J}_{ijkl}, \tag{20}$$

$$
\begin{aligned}
{}^{0}\mathbb{C}^{\nabla Jdamp} &= 2G^{damp}\mathbb{I} + \lambda^{damp}(\mathbf{1}\otimes\mathbf{1}),\ G^{damp} = \frac{E^{damp}}{2(1+\nu^{damp})},\\
\lambda^{damp} &= \frac{\nu^{damp}E^{damp}}{(1+\nu^{damp})(1-2\nu^{damp})},
\end{aligned}
\tag{21}
$$

$$
{}^{n+1}\hat{\mathbb{C}}^{\nabla Jdamp}_{abcd} = \frac{1}{{}^{n+1}J}\,{}^{n+1}R_{ma}\,{}^{n+1}R_{nb}\,{}^{n+1}R_{oc}\,{}^{n+1}R_{pd}\,{}^{n+1}F_{mi}\,{}^{n+1}F_{nj}\,{}^{n+1}F_{ok}\,{}^{n+1}F_{pl}\,{}^{0}\mathbb{C}^{\nabla Jdamp}_{ijkl},
\tag{22}
$$

In the co-rotational elastic Cauchy's stress integration, Eq. (15), the midpoint rule and the Jaumann's objective rate is used, which can also be expressed in an equivalent Green-Naghdi's objective rate form, if the corresponding rotation tensors take the following forms [26, 84].

$$
{}^{n+\frac{1}{2}}\mathbf{R} = \exp\left[\frac{\Delta t}{2}\,{}^{n+\frac{1}{2}}\mathbf{W}\right]\cdot{}^{n}\mathbf{R},
\tag{23}
$$

$$
{}^{n+1}\mathbf{R} = \exp\left[\Delta t^{n+\frac{1}{2}}\mathbf{W}\right]\cdot{}^{n}\mathbf{R},
\tag{24}
$$

where $\mathbf{W}$ stands for the spin tensor. In the tensor exponential function (23) and (24) calculation the Rodrugues' formula [84] was used. No objective integration is needed in the co-rotational viscous Cauchy's stress tensor calculation (18), as the stress tensor is objective and is instantaneous in its nature. In Eqs. (14)–(24), (32)–(37) the left superscripts n, $n+1/2$, $n+1$ denote the physical quantity value at discrete times, corresponding to previous, mid and current configuration of the body within the current time step Δt. We assume that the material in the initial configuration of the body is isotropic, and that the co-rotational strain rate tensor $\hat{\mathbf{d}} = \mathbf{R}^T\cdot\mathbf{d}\cdot\mathbf{R}$ has the additive decomposition $\hat{\mathbf{d}} = \hat{\mathbf{d}}^{el} + \hat{\mathbf{d}}^{pl} + \hat{\mathbf{d}}^{th}$ into an elastic part $\hat{\mathbf{d}}^{el}$, a plastic part $\hat{\mathbf{d}}^{pl}$ and a thermal part $\hat{\mathbf{d}}^{th} = \alpha\dot{T}\mathbf{R}^T\cdot\mathbf{R} = \alpha\dot{T}\mathbf{I}$. Here α is the coefficient of thermal expansion, $\dot{T}$ is the temperature change per unit time and $\mathbf{R}$ stands for the second-order rotation tensor (23), (24). In this work the St. Venant-Kirchov's material model is used [85], the elastic material tensor ${}^{0}\mathbb{C}^{\nabla J}$ of which can be expressed with Eq. (19) in the initial configuration. Here G and λ denote the shear modulus and the Lame's constant, $\mathbb{I}, \mathbf{I}$ stand for a fourth-order unit tensor and a second-order unit tensor and ν, E are respectively the Poisson's ratio and the Young's modulus. The fourth-order damping tensor ${}^{0}\mathbb{C}^{\nabla Jdamp}$ is formally constructed in the same way, Eq. (21) in the initial configuration as the elastic material tensor using two independent variables ν^{damp}, E^{damp}, which ensures isotropy. The advantage of such a formulation is that it enables easy experimental identification of the material parameters using amplitude measurements of thin-walled tubes in free, longitudinal and torsional vibration. In order to use the fourth-order elastic and damping material tensor for stress update calculation, they have to be transformed into the mid and current configuration within the current time step and subsequently rotated into the reference configuration for stress update calculation [82] using Eqs. (20) and (22), in which $\mathbf{F}$ stands for the deformation gradient and $J = \det(\mathbf{F})$ is the Jacobian or the determinant of $\mathbf{F}$ respectively. Equations (14)–(24) are supplemented with the following constitutive and evolution equations:

$$f = \sigma_{eq} - \sigma_y - R \leq 0, \tag{25}$$

$$\sigma_{eq} = \sqrt{\frac{3}{2}\left(\hat{\Sigma} - \Sigma\hat{\mathbf{X}}\right) : \left(\hat{\Sigma} - \Sigma\hat{\mathbf{X}}\right)}, \tag{26}$$

$$R = Q\left(1 - e^{-b\varepsilon^{pl}}\right), \tag{27}$$

$$\dot{\mathbf{X}}^{\nabla J} = \mathbb{C}^{\nabla Jcycl} : \mathbf{d}^{pl} - \gamma\left(\varepsilon^{pl}\right)\mathbf{X}\dot{\varepsilon}^{pl}, \quad tr(\mathbf{X}) \neq 0, \tag{28}$$

$$\gamma\left(\varepsilon^{pl}\right) = \gamma_{\infty} - (\gamma_{\infty} - \gamma_0)e^{\left(-\omega\varepsilon^{pl}\right)}, \tag{29}$$

$$\begin{aligned} {}^0\mathbb{C}^{\nabla Jcycl} &= 2G^{cycl} + \lambda^{cycl}(\mathbf{1} \otimes \mathbf{1}), \quad G^{cycl} = \frac{E^{cycl}}{2(1 + \nu^{cycl})}, \\ \lambda^{cycl} &= \frac{\nu^{cycl}E^{cycl}}{(1 + \nu^{cycl})(1 - 2\nu^{cycl})}, \end{aligned} \tag{30}$$

$$\mathbb{C}^{\nabla Jcycl}_{mnop} = \frac{1}{J} \cdot F_{mi} \cdot F_{nj} \cdot F_{ok} \cdot F_{pl} \cdot {}^0\mathbb{C}^{\nabla Jcycl}_{ijkl}, \tag{31}$$

$$ {}^{n+1}\hat{\mathbf{X}} = \Delta t \cdot \left[{}^{n+\frac{1}{2}}\hat{\mathbb{C}}^{\nabla Jcycl} : {}^{n+\frac{1}{2}}\hat{\mathbf{d}}^{pl} - \gamma\left({}^{n+\frac{1}{2}}\varepsilon^{pl}\right){}^{n+\frac{1}{2}}\hat{\mathbf{X}}{}^{n+\frac{1}{2}}\dot{\varepsilon}^{pl}\right] + {}^{n}\hat{\mathbf{X}}, \; tr\left({}^{n+1}\hat{\mathbf{X}}\right) \neq 0, \tag{32}$$

$$ {}^{n+\frac{1}{2}}\hat{\mathbb{C}}^{\nabla Jcycl}_{abcd} = \frac{1}{{}^{n+\frac{1}{2}}J}{}^{n+\frac{1}{2}}R_{ma}{}^{n+\frac{1}{2}}R_{nb}{}^{n+\frac{1}{2}}R_{oc}{}^{n+\frac{1}{2}}R_{pd}{}^{n+\frac{1}{2}}F_{mi}{}^{n+\frac{1}{2}}F_{nj}{}^{n+\frac{1}{2}}F_{ok}{}^{n+\frac{1}{2}}F_{pl}{}^{0}\mathbb{C}^{\nabla Jcycl}_{ijkl}, \tag{33}$$

$$\Delta t \cdot {}^{n+\frac{1}{2}}\hat{\mathbf{d}}^{pl} = \Delta\tilde{\lambda} \cdot \sqrt{\frac{3}{2}} \cdot \hat{\mathbf{N}}, \quad \hat{\mathbf{N}} = \frac{{}^{n+1}\hat{\Sigma} - {}^{n+1}\Sigma\hat{\mathbf{X}}}{\sqrt{\left({}^{n+1}\hat{\Sigma} - {}^{n+1}\Sigma\hat{\mathbf{X}}\right) : \left({}^{n+1}\hat{\Sigma} - {}^{n+1}\Sigma\hat{\mathbf{X}}\right)}}, \tag{34}$$

$$\Delta\tilde{\lambda} = \frac{\sqrt{\frac{3}{2}}\hat{\mathbf{N}} : \left[{}^{n+\frac{1}{2}}\hat{\mathbb{C}}^{\nabla J} : \left(\Delta t^{n+\frac{1}{2}}\hat{\mathbf{d}} - \Delta t^{n+\frac{1}{2}}\hat{\mathbf{d}}th\right) + {}^{n+1}\hat{\mathbb{C}}^{\nabla Jdamp} : {}^{n+1}\hat{\mathbf{d}} - {}^{n}\hat{\mathbb{C}}^{\nabla Jdamp} : {}^{n}\hat{\mathbf{d}}\right]}{\frac{3}{2}\hat{\mathbf{N}} : {}^{n+\frac{1}{2}}\hat{\mathbb{C}}^{\nabla J} : \hat{\mathbf{N}} + \frac{3}{2}\hat{\mathbf{N}} : {}^{n+\frac{1}{2}}\hat{\mathbb{C}}^{\nabla Jcycl} : \hat{\mathbf{N}} - \sqrt{\frac{3}{2}}\gamma\left({}^{n+\frac{1}{2}}\varepsilon^{pl}\right)\hat{\mathbf{N}} : {}^{n+\frac{1}{2}}\hat{\mathbf{X}} + Qbe^{-b\left(\Delta\tilde{\lambda} + {}^{n}\varepsilon^{pl}\right)}}, \tag{35}$$

where

$$\gamma\left({}^{n+\frac{1}{2}}\varepsilon^{pl}\right) = \gamma_\infty - (\gamma_\infty - \gamma_0)e^{-\omega\left(\frac{\Delta\tilde{\lambda}}{2}+{}^n\varepsilon^{pl}\right)}, \tag{36}$$

$${}^n\hat{\mathbb{C}}^{\nabla Jdamp}_{abcd} = \frac{1}{{}^nJ}{}^nR_{ma}{}^nR_{nb}{}^nR_{oc}{}^nR_{pd}{}^nF_{mi}{}^nF_{nj}{}^nF_{ok}{}^nF_{pl}{}^0\mathbb{C}^{\nabla Jdamp}_{ijkl}. \tag{37}$$

Equations (25)–(31) formulate the yield surface and the evolution equations of the extended NoIHKH material model [86] for cyclic plasticity of metals. The NoIH rule for isotropic hardening and the NoKH rule for kinematic hardening are given with Eqs. (27) and (28), where the later can be manipulated in the form of Eq. (32) after objective integration using the midpoint rule and the Jauman's objective rate in the form of the Green-Naghdi's objective rate. Here $\hat{\Sigma}, \mathbf{X}, \hat{\mathbf{X}}, \Sigma\hat{\mathbf{X}}, \varepsilon^{pl}, \dot{\varepsilon}^{pl}$ are the deviatoric component of the co-rotational Cauchy stress tensor, the back stress tensor, the co-rotational back stress tensor $\hat{\mathbf{X}} = \mathbf{R}^T \cdot \mathbf{X} \cdot \mathbf{R}$, the deviatoric component of the co-rotational back stress tensor, the accumulated plastic strain and the accumulated plastic strain rate respectively. The remaining symbols denote constant material parameters. The fourth order cyclic material tensor ${}^0\mathbb{C}^{\nabla Jcycl}$ in the initial configuration is formally constructed in the same way as the elastic material tensor using two independent variables ν^{cycl}, E^{cycl}, which ensure isotropy. In order to use the tensor for stress update calculation, it has to be transformed into the mid configuration of the body within the current time step and subsequently rotated into the reference configuration using Eq. (33). Furthermore, a slight modification is needed in the effective stress definition (26), where the back stress has to be replaced with its deviatoric component, as the transformation (33) results in the co-rotational back stress tensor (32) being non-deviatoric. This property, namely the back stress tensor being deviatoric, is explicitly required in contemporary J_2 flow theories and is assured by the choice of an appropriate evolution equation [26]. Since the rotation tensor (24) is a mixed tensor, which transforms a tensor field expressed in terms of an initial basis into a tensor field expressed in terms of a current basis, in the co-rotational stress update calculation (14) and (32) all tensors are associated with a fixed co-ordinate system, parallel to the initial co-ordinate system of the undeformed body, which formally makes the stress update similar to the one used in small strain elastoplasticity. Using the principle of maximum plastic dissipation and associative plasticity, equations for the plastic strain increment and the plastic multiplier can be derived in the form (34) and (35), where $\hat{\mathbf{N}}$ is a second order tensor defining the normal direction of the yield surface. During plastic loading Eqs. (35), (32) and (14) are solved for unknowns $\Delta\tilde{\lambda}, {}^{n+1}\hat{\mathbf{X}}, {}^{n+1}\hat{\sigma}$, where the later two tensors are subsequently rotated back into the current configuration for further usage. When crossing the yield surface no special integration procedure is used to locate the intersection of the elastic stress tensor with the yield surface [87], which in repeated crossing during cyclic loading leads to error accumulation.

2.4 Dissipation Induced Heating

During deformation a certain amount of the supplied mechanical energy either dissipates into heat or gets consumed by other dissipation processes, such as creation and rearrangement of crystal defects and formation of dislocation structures [90]. To determine the exact portion of the dissipated energy that changes into heat represents a difficult experimental task [88, 89]. A typical ratio of plastic work converted into heat is between 0.9 and 1.0 [16], however this value might greatly differ for various steels with average values between 0.75 and 1.0 [90]. The ratio is generally assumed to be constant, independent of plastic deformation and strain rate. In this paper, we assume that this figure is around 80 % and then we define the heat generation rate per unit volume as follows

$$r = 0.8\boldsymbol{\sigma}^{el} : \mathbf{d}^{pl} + 0.8\boldsymbol{\sigma}^{damp} : \mathbf{d}. \tag{38}$$

Here the first term on the right-hand side denotes the plastic heating and the second term stands for the internal damping induced heating.

2.5 Phase Transformation Equations

When the temperature of the body at some material point reaches the recrystallization temperature, phase changes take place in the material and this results in the austenite phase to turn into a ferrite, a perlite, a bainite and a martensite phase in cooling or vice versa in heating. A typical analytical approach to determine the phase content in the material during recrystallization use the Johnson and Mehl [91] and Avrami [92] equation for ferrite, perlite, bainite transformation under isothermal condition together with the Koistinen and Marburger equation [93] for an austenite-martensite transformation, the material parameters of which can be easily determined from a continuous cooling transformation (CCT) diagram of the material. Figure 1 shows a CCT diagram of a low carbon steel, where the aforementioned phase regions, recrystallization temperatures and transformation start and end times can be identified. Although at present other more sophisticated models can be employed, which take into account the transformation-induced plasticity too [94, 95] in the material, in this paper for simplicity we will still use the aforementioned analytical approach to determine the phase content. The simplified rate form of the Mehl-Avrami transformation equation is then used to determine the ferrite, the perlite and the bainite phase content in the analysis when cooling takes place [96], which was derived from the Mehl-Avrami equation, defined with the following formula.

$$y_i(T,t) = y_{ieq}(T)\left\{1 - \exp\left[-b_i(T)(t - t_{cr})^{n_i(T)}\right]\right\}, \tag{39}$$

where the index $i = 2, 3, 4$ stands for the ferrite, the pearlite and the bainite phase, $y_{ieq}(T)$ denotes an equivalent ith phase volume fraction, which represents the maximum volume fraction of the given phase that can be created during the transformation process at a given $T = const$ temperature, T, t, t_{cr} stand for a material temperature, a phase transformation time and a starting time of the cooling/heating cycle. The material properties $b_i(T)$, $n_i(T)$ can be calculated after identifying the transformation start and end times from the CCT diagram, using the procedure described in [97].

A similar rate equation is used to determine the martensite content in the material when cooling takes place and it is analogically derived from the Koisten-Marburger transformation equation

$$y_5(T) = y_{5eq}(T)\{1 - \exp[-\alpha(M_s - T)]\}, \tag{40}$$

where α, M_s are material constants, which can be similarly calculated using the CCT diagram. The corresponding simplified rate forms then can be expressed with the following formulas:

$$\dot{y}_i = (y_{ieq} - y_i)b_i n_i \left[\frac{-\ln\left(\frac{y_i - y_{ieq}}{y_{ic} - y_{ieq}}\right)}{b_i}\right]^{\frac{n_i - 1}{n_i}}, \quad \text{for} \quad i = 2, 3, 4 \tag{41}$$

$$n_i(T) = \frac{\ln\left(\frac{\ln(0.99)}{\ln(0.01)}\right)}{\ln\left(\frac{t_{is}(T) - t_{cr}}{t_{if}(T) - t_{cr}}\right)} = \frac{\ln\left(\frac{\ln(0.99)}{\ln(0.01)}\right)}{\ln\left(\frac{\hat{t}_{is}(T)}{\hat{t}_{if}(T)}\right)}, \tag{42}$$

$$b_i(T) = \frac{-\ln(0.01)}{(t_{if} - t_{cr})^{n_i(T)}} = \frac{-\ln(0.01)}{(\hat{t}_{if})^{n_i(T)}} = \frac{-\ln(0.01)}{(\hat{t}_{if})^{\frac{\ln\left(\frac{\ln(0.99)}{\ln(0.01)}\right)}{\ln\left(\frac{\hat{t}_{is}(T)}{\hat{t}_{if}(T)}\right)}}}, \tag{43}$$

where $t_{if} = \widehat{t}_{if} + t_{cr}$, $t_{is} = \widehat{t}_{is} + t_{cr}$ and $\widehat{t}_{is}, \widehat{t}_{if}, y_{ic}$ are the ith phase transformation start and end times corresponding to 1/99 % of their volume fraction, and the ith phase volume fraction at time t_{cr} (See Fig. 1, where $t_{cr} = 0$).

$$\dot{y}_5 = \left(y_{5eq} - y_5\right)\alpha\dot{T}. \tag{44}$$

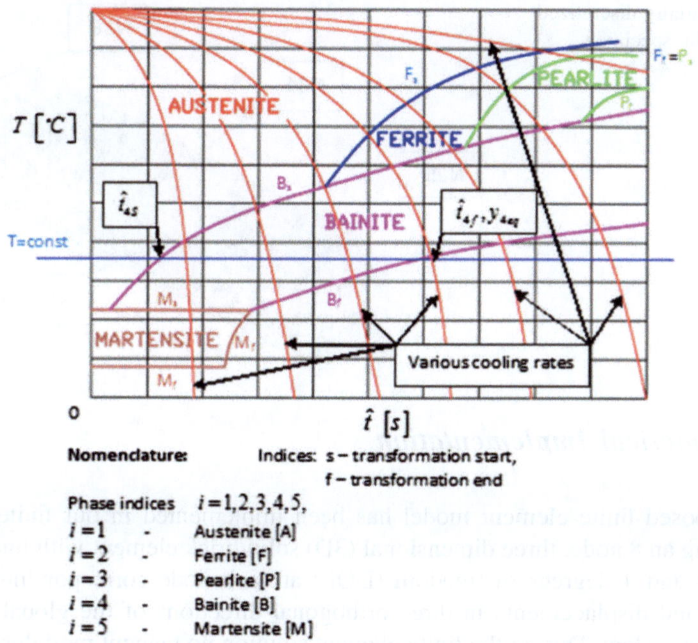

Fig. 1 Continuous cooling transformation (*CCT*) diagram

If the material temperature increases, reverse phase transformations take place in the metal above its recrystallization temperature. To model these situations the authors modied the Mehl-Avrami and the Koisten-Marburger equations, which resulted in the following formulas for the phase transformations

$$y_i(T,t) = y_{ieq}(T)\exp\left[-b_i(T)(t-t_{cr})^{n_i(T)}\right], \quad \text{for} \quad i = 2,3,4 \tag{45}$$

$$y_5(T) = y_{5eq}(T)\exp[-\alpha(M_s - T)]. \tag{46}$$

The corresponding simplified rate forms then can be expressed as follows using the same procedure as in the case of cooling:

$$\dot{y}_i = -y_i b_i n_i \left[\frac{-\ln\left(\frac{y_i}{y_{ieq}}\right)}{b_i}\right]^{\frac{n_i-1}{n_i}}, \quad \text{for} \quad i = 2,3,4 \tag{47}$$

$$\dot{y}_5 = y_5 \alpha \dot{T}. \tag{48}$$

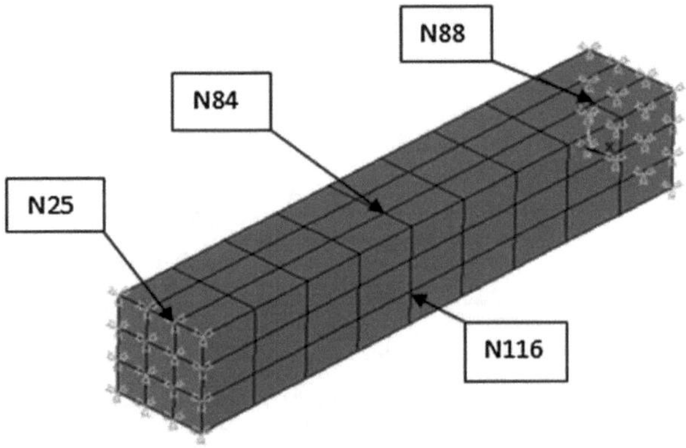

Fig. 2 Spatially discretized model of the specimen

2.6 Numerical Implementation

The proposed finite element model has been implemented in our finite element code using an 8 node, three dimensional (3D) solid/brick element with linear shape functions and 4 degrees of freedom (DOF) at each node corresponding to temperature and displacements in three orthogonal directions of the global Eulerian coordinate system. During the finite element solution no tangent modulus has been calculated, only the system of algebraic equations corresponding to local extremes of functionals (6) and (7) obtained after their derivation according to the virtual degrees of freedom has been solved for unknown DOF using the Newton's method. The analysis has been run as transient dynamic using the trapezoidal rule in implicit time integration. The phase volume fraction of the ferrite, pearlite, bainite, martensite and the retaining austenite phase has been calculated as a secondary variable at Gauss' points of each finite element. In order to use real material properties a 64 bit finite element code has been generated and run on a 64 bit PC to have 33 valid digits after a decimal point in the analysis.

3 Numerical Example

As a numerical example a straight bar, size 10 mm × 10 mm × 60 mm loaded in cyclic bending is studied. One end of the bar is fixed and the other end undergoes a prescribed vertical (Y) deformation defined by a sine function and linearly changing amplitude from zero to a maximum in the run-up stage until the maximum, 4 mm amplitude is reached in the operation stage. The moving end of the specimen is guided, with all degrees of freedom taken in the remaining (X and Z) directions at each node located on the end. The velocity and the acceleration of the moving end of the bar are defined as first and second derivatives of the prescribed vertical deformation. Figure 2 shows the spatially discretized model of the specimen. In the experiment 17.5 loading cycles are realized using 1 Hz loading

Table 1 Material properties

ρ (kg m^{-3})	7,800.0
E (Pa)	2.1×10^{11}
E^{cycl} (Pa)	2.1×10^{5}
E^{damp} (Pa s)	2.1×10^{7}
$\nu = \nu^{cycl} = \nu^{damp}$ (–)	0.3
$\alpha_x = \alpha_y = \alpha_z = \alpha$ (°C^{-1})	0.0000126
c (J/kg °C)	500.0
$k_{xx} = k_{yy} = k_{zz}$ (W/m °C)	45.0
h (W/m^2 °C)	500.0

Table 2 Hardening properties

σ_y (Pa)	350.0×10^6
Q (Pa)	150.0×10^6
b (–)	3.0
γ_∞ (–)	0.001
γ_0 (–)	0.002
ω (–)	10.0

Table 3 Ferrite phase transformation material properties

T (°C)	y_{2eq} (–)	F_s (s)	F_f (s)
580	0.05	21.0	100.0
600	0.07	18.0	101.0
630	0.18	20.0	150.0
670	0.30	25.0	400.0
700	0.40	55.0	2,000.0
740	0.40	400.0	100,000.0
770	0.40	10,000	1,000,000.0

Table 4 Perlite phase transformation material properties

T (°C)	y_{3eq} (–)	P_s (s)	P_f (s)
510	0.10	580.0	600.0
550	0.15	250.0	500.0
580	0.30	100.0	480.0
600	0.40	101.0	420.0
630	0.60	150.0	400.0
650	0.70	200.0	500.1
680	0.60	380.0	3,000.0
700	0.60	2,000.0	100,000.0

frequency, from which the run-up takes 10 cycles. The analysis has been run as transient-dynamic using $1.041\bar{6} \times 10^{-2}$ s time step size. Heat convection through all external surfaces is considered, applying zero bulk/environmental temperature. The bar is initially at rest with 770 °C initial temperature, well above the austenite recrystallization temperature, which means that at the beginning only the austenite phase is present in the material.

Table 5 Bainite phase transformation material properties

T (°C)	y_{4eq} (–)	B_s (s)	B_f (s)
280	1.0	1,200.0	1,200.001
300	0.8	500.0	1,100.0
370	0.7	6.0	1,000.0
400	0.6	5.0	900.0
450	0.4	5.001	700.0
500	0.2	6.001	600.0
550	0.1	10.0	250.0
580	0.05	20.0	150.0

Table 6 Martensite phase transformation material properties

T (°C)	y_{5eq} (–)	M_s (s)	M_f (s)	α
260	1.0	0.0	1,200.0	0.0177121
270	0.95	0.0	1,000.0	0.0170561
280	0.85	0.0	600.0	0.0164470
320	0.75	0.0	120.0	0.0143911
340	0.03	0.0	30.0	0.0135446
360	0.02	0.0	12.0	0.0127921

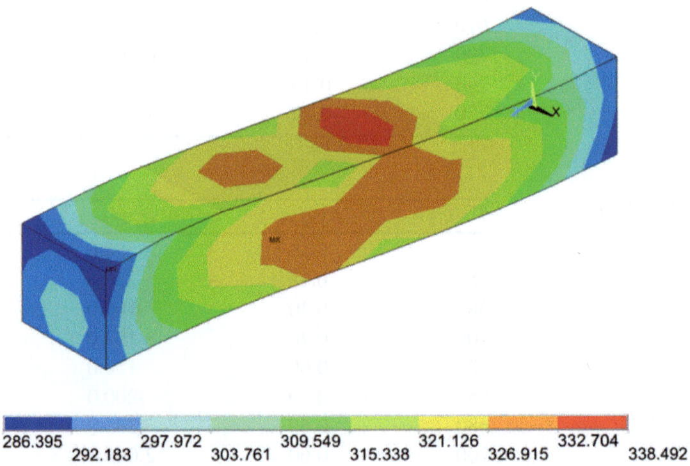

Fig. 3 Temperature distribution at maximum deformation in the last cycle

In the numerical experiment constant material properties of low carbon steel are used disregarding the latent heat during phase transformation. These restrictions are the results of the present state of the used finite element code, in which the appropriate modules have not yet been implemented. Tables 1, 2, 3, 4, 5, 6 list the used material properties.

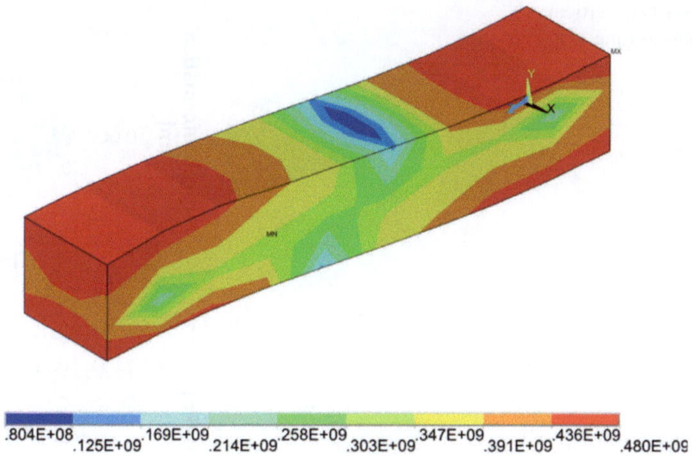

Fig. 4 Von mises stress distribution at maximum deformation in the last cycle

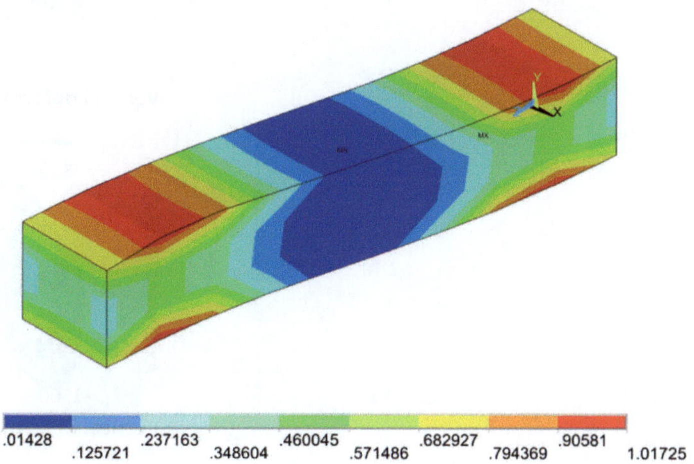

Fig. 5 Accumulated plastic strain distribution at maximum deformation in the last cycle

4 Numerical Results

Figures 3, 4, 5, 6, 7, 8, 9, 10, 11, 12, 13, 14 show some selected results of the finite element analysis. They are the temperature distribution, the Von-Mises stress distribution and the accumulated plastic strain distribution at maximum vertical deformation of the moving end of the bar in the last loading cycle (Figs. 3, 4, 5). The prescribed deformation time history at node 25, the created hysteresis loops at nodes 25 and 88 and the temperature time history curves at nodes 25, 84, 88 and 116 are shown in Figs. 6, 7, 8 (see Fig. 2 for the exact location of the nodes).

Fig. 6 Prescribed vertical (Y) deformation time history

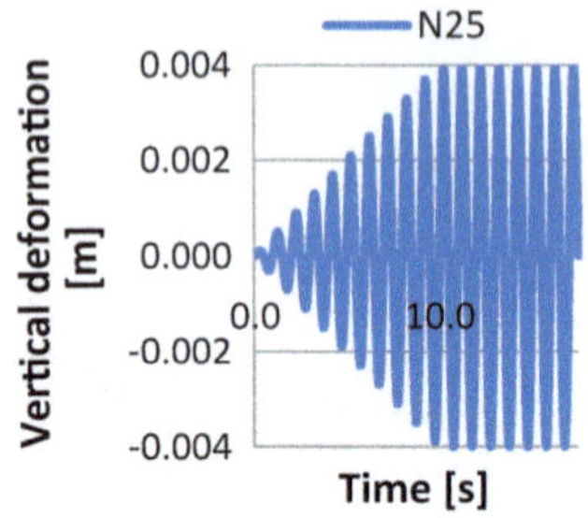

Fig. 7 Hysteresis loop development at nodes 25 and 88

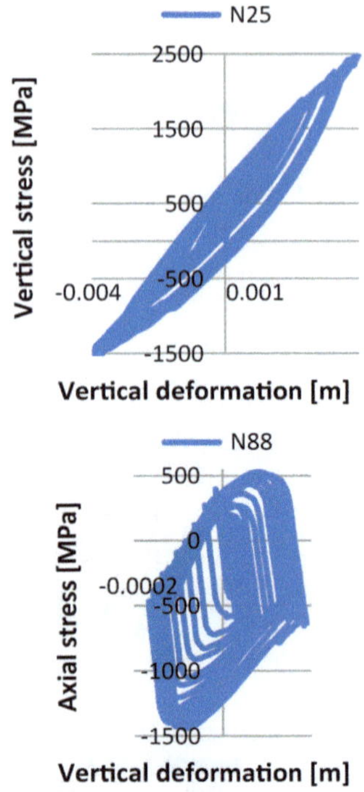

Fig. 8 Temperature time history at nodes 25, 84, 88 and 116

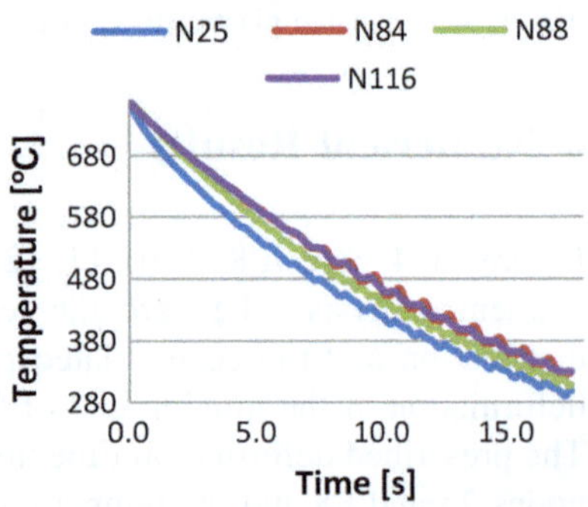

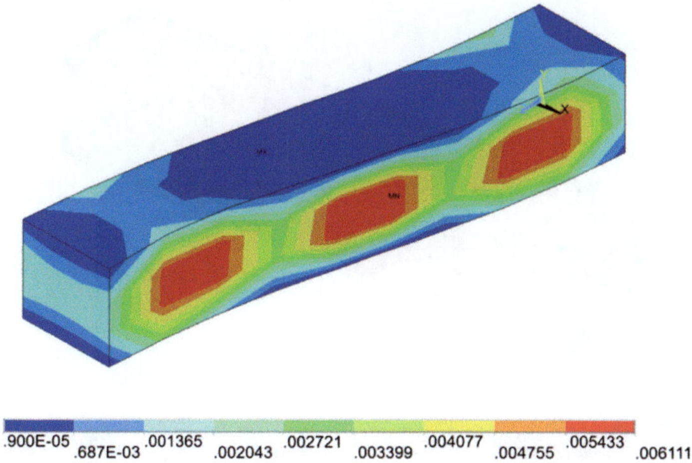

Fig. 9 Bainite volume content at maximum deformation in the last cycle

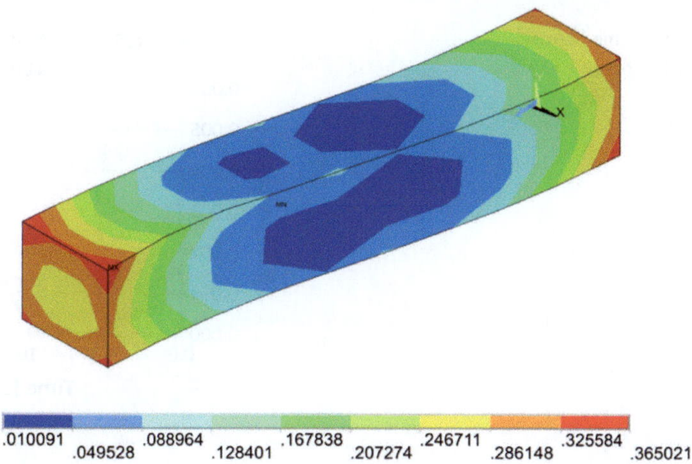

Fig. 10 Martensite volume content at maximum deformation in the last cycle

These figures show that that the cooling results in the hysteresis loops are not symmetric. Furthermore temperature oscillation can be observed in the temperature time history curves as a result of the dissipation induced heating. Figures 9, 10, 11 depict the bainite, martensite and the retaining austenite volume content in the last loading cycle corresponding to the maximum prescribed vertical deformation of the moving end of the bar. In the numerical analysis bainite and martensite phases have been created only with no feerite and pearlite phase transformations as the length of the cooling period at a material point has not allowed growing the aforementioned phases. In order to follow more closely the phase transformations in the material bainite, martensite and austenite time history

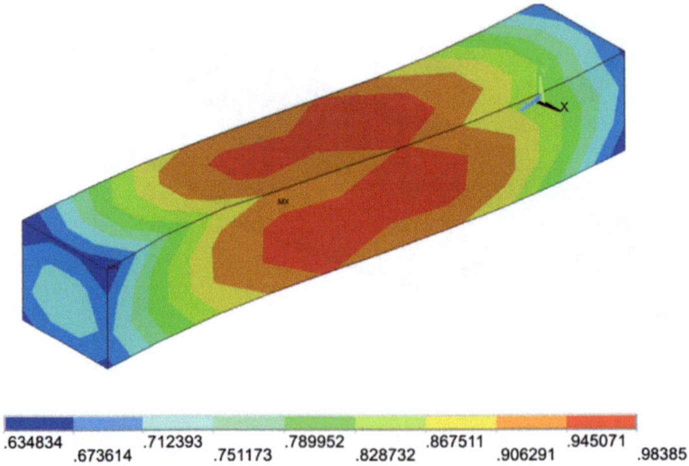

Fig. 11 Austenite volume content at maximum deformation in the last cycle

Fig. 12 Bainite time history at nodes 25, 84, 88 and 116

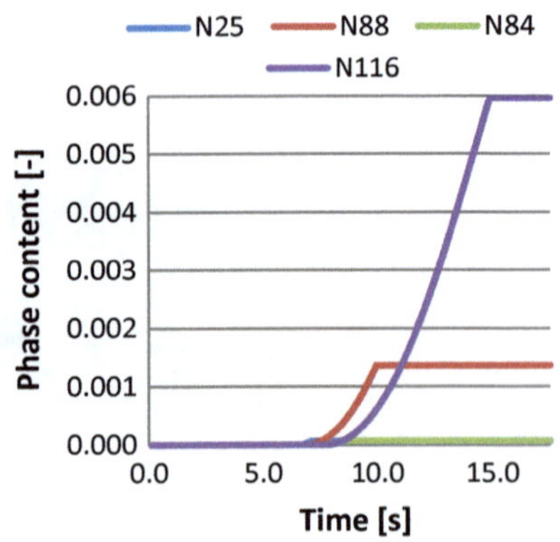

Fig. 13 Martensite time history at nodes 25, 84, 88 and 116

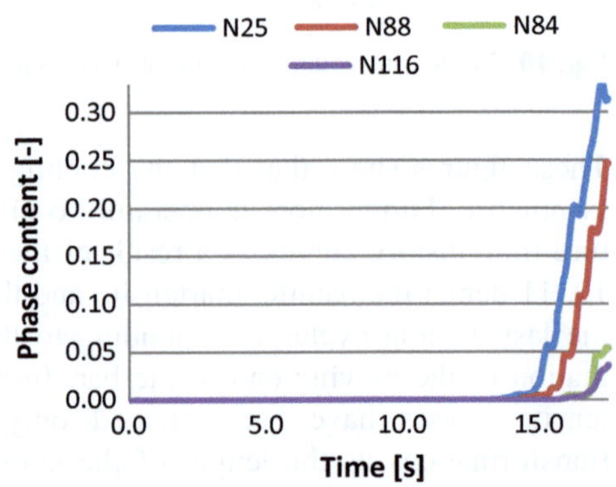

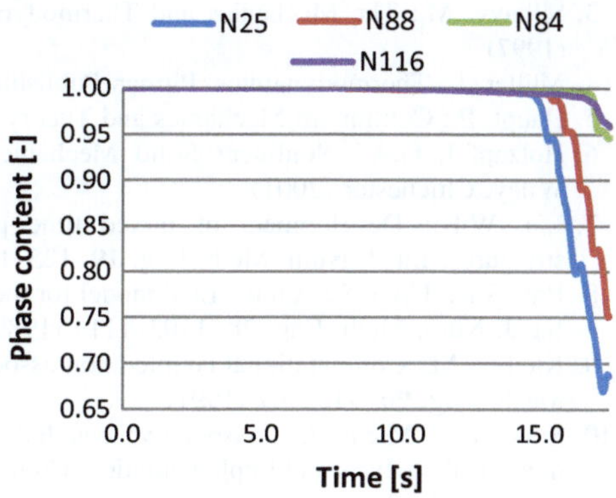

Fig. 14 Austenite time history at nodes 25, 84, 88 and 116

curves have been created (Figs. 12, 13, 14). Again we can see the phase volume content fluctuations in the figures, i.e. the austenite transforms into martensite and vice versa, due to the presence of the repeated cooling heating periods, which is the consequence of the dissipation induced heating that originates from mechanical loading.

5 Conclusive Summary

In this work recent development in the solution of fully coupled thermal structural problems with convective heat transfer and phase changes has been presented. The calculation is based on the improved weak form for the updated Lagrangian formulation, which is capable of simulating iron construction behaviour under extreme conditions that include large strains/large deformations at high strain rates and temperatures, employing the extended NoIHKH material model for cyclic plasticity of metals, the Kelvin–Voigt model for internal damping and the Mehl-Avrami and Koisten-Marburger phase transformation equations. The selected results represent some of the first outcomes of the problem finite element solution. Although the proposed model still needs to be experimentally verified, the authors believe that it creates new perspectives in the numerical study of real materials.

Acknowledgments Funding using the VEGA grant 1/0488/09 and 1/0051/10 resources is greatly appreciated.

References

1. Skrzypek, J.J., Ganczarski, A.W., Rustichelli, F., Egner, H.: Advanced Materials and Structures for Extreme Operating Conditions. Springer, Berlin (2008)
2. Maugin, G.A.: The Thermomechanics of Plasticity and Fracture. Cambridge University Press, Cambridge (1992)

3. Šilhavý, M.: The Mechanics and Thermodynamics of Continuous Media. Springer, Berlin (1997)
4. Müller, I.: Thermodynamics. Pitman Publishing LTD, London (1985)
5. Haupt, P.: Continuum Mechanics and Theory of Materials, 2nd edn. Springer, Berlin (2002)
6. Holzapfel, G.A.: Nonlinear Solid Mechanics. A Continuum Approach for Engineering. Wiley, Chichester (2001)
7. Liu, W.K.: Development of mixed time partition procedures for thermal analysis of structures. Int. J. Num. Meth. Eng. **19**, 125–140 (1983)
8. Ray, S.K., Utku, S.: A numerical model for the thermo-elasto-plastic behaviour of a material. Int. J. Num. Meth. Eng. **28**, 1103–1114 (1989)
9. Kleiber, M.: Computational coupled non-associative thermo-plasticity. Comput. Meth. Appl. Mech. Eng. **90**, 943–967 (1991)
10. Simo, J.C., Miehe, C.: Associative coupled thermoplasticity at finite strains: Formulation, numerical analysis and implementation. Comput. Meth. Appl. Mech. Eng. **98**, 41–104 (1992)
11. Služalec, A.: Temperature rise in elastic-plastic metal. Comput. Meth. Appl. Mech. Eng. **96**, 293–302 (1992)
12. Huang, Z.P.: Arate independent thermoplastic theory at finite deformations. Arch. Mech. **46**(6), 855–879 (1994)
13. Pantuso, D., Bathe, K.J., Bouzinov, P.A.: A finite element procedure for the analysis of thermo-mechanical solids in contact. Comput. Struct. **75**, 551–573 (2000)
14. Batra, R.C., Love, B.M.: Mesoscale analysis of shear bands in high strain rate deformations of tungsten/nickel-iron composites. J. Therm. Stresses **28**, 747–782 (2005)
15. Maugin, A.G., Berezovski, A.: On the propagation of singular surfaces in thermoelasticity. J. Therm. Stresses **32**, 557–592 (2009)
16. Služalec, A.: An evaluation of the internal dissipation factor in coupled thermoplasticity. Int. J. Non-Lin. Mech. **25**(4), 395–403 (1990)
17. Dillon Jr, O.W.: Coupled thermoplasticity. J. Mech. Phys. Solids **11**, 21–33 (1963)
18. Saracibar, C.A., Cervena, M., Chiumenti, M.: On the constitutive modeling of coupled thermomechanical phase-change problems. Int. J. Plast. **17**, 1565–1622 (2001)
19. Júnior, M.V: Computational approaches to simulation of metal cutting processes. Thesis, University of Wales. Swansea (1998)
20. Schönauer, M.: Unified numerical analysis of cold and hot metal forming processes. Thesis, University of Wales, Swansea (1993)
21. Oden, T.J.: Finite Elements of Nonlinear Continua. McGraw-Hill, New York (1972)
22. Lemaitre, J., Chaboche, J.L.: Mechanics of Solid Materials. Cambridge University Press, Cambridge (1994)
23. Ibrahimbegovic, A.: Nonlinear Solid Mechanics. Theoretical Formulations and Finite Element Solution Methods. Springer, Dordrecht (2009)
24. Écsi, L., Élesztős, P.: Constitutive equation with internal damping for materials under cyclic and dynamic loadings using a fully coupled thermal-structural finite element analysis. Int. J. Multiphys. **3**(2), 155–165 (2009)
25. Porter, D.A., Easterling, K.E.: Phase Transformations in Metals and Alloys, 2nd edn. Chapman & Hall, London (1992)
26. Simo, J.C., Hughes, T.J.R.: Computational Inelasticity. Springer, New York (1998)
27. Nemat-Nasser, S.: Plasticity. A Treatsie on Finite Deformation of Heterogenous Inelastic Materials. Cambridge University Press, Cambridge (2009)
28. Lemaitre, J.: Handbook of Material Behavior Models. Deformations of Materials, vol. 1. Academic Press, London (2001)
29. Lemaitre, J.: Handbook of Material Behavior Models. Failures of Materials, vol. 2. Academic Press, London (2001)
30. Lemaitre, J.: Handbook of Material Behavior Models. Multiphysics Behaviors, vol. 3. Academic Press, London (2001)
31. Suresh, S.: Fatigue of Materials, 2nd edn. Cambridge University Press, Cambridge (1998)
32. Lemaitre, J.: A Course on Damage Mechanics. Springer, Berlin (1992)

33. Needleman, A., Tvergaard, V.: Analysis of a brittle-ductile transition under dynamic shear loading. Int. J. Solids Struct. **32**(17–18), 2571–2590 (1995)
34. Zhou, M., Ravichandran, G., Rosakis, A.J.: Dynamically propagating shear bands in impact-loaded prenotched plates—II. Numerical simulations. J. Mech. Phys. Solids **44**(6), 1007–1032 (1996)
35. Li, S., Liu, W.K., Rosakis, A.J., Belytschko, T., Hao, V.: Mesh-Free Galerkin simulations of dynamic shear band propagation and failure mode transition. Int. J. Solids Struct. **39**, 1213–1240 (2002)
36. Liang, R., Khan, A.S.: A critical review of experimental results and constitutive models for BCC and FCC metals over wide range of strain rates and temperatures. Int. J. Plast. **15**, 963–980 (1999)
37. Caboche, J.L.: A review of some plasticity and viscoplasticity constitutive theories. Int. J. Plast. **24**, 1642–1693 (2008)
38. Rusinek, A., Rodríguez-Martínez, J.A., Arias, A.: A thermo-viscoplastic model for FCC metals with application to OFHC copper. Int. J. Mech. Sci. **52**, 120–135 (2010)
39. Holmquist, T.J., Templeton, D.W., Bishnoi, K.D.: Constitutive modelling of aluminium nitride for large strain high-strain rate, and high-pressure applications. Int. J. Impact Eng. **25**, 211–231 (2001)
40. Rodríguez-Martínez, J.A., Rusinek, A., Klepaczko, J.R.: Constitutive relation for steels approximating quasi-static and intermediate strain rates at large deformations. Mech. Res. Commun. **36**, 419–427 (2009)
41. Yu, H., Guo, Y., Zhang, K., Lai, X.: Constitutive model on the description of plastic behavior of DP600 steel at strain rate from 10^{-4} to 10^{3} s^{-1}. Comput. Mater. Sci. **46**, 36–41 (2009)
42. Rusinek, A., Zaera, R., Klepaczko, J.R.: Constitutive relation in 3-D for wide range of strain rates and temperatures-application to mild steels. Int. J. Solid. Struct. **44**, 5611–5634 (2007)
43. Yu, H., Guo, Y., Lai, X.: Rate-dependent behavior and constitutive model of DP600 steel at strain rate from 10^{-4} to 10^{3} s^{-1}. Mater. Des. **30**, 2501–2505 (2009)
44. Yin, Z.N., Wang, T.J.: Deformation of PC/ABS alloys at elevated temperatures and high strain rates. Mat. Sci. Eng. A **494**, 304–313 (2008)
45. Deseri, L., Mares, R.: A class of viscoelastoplastic constitutive models based on the maximum dissipation principle. Mech. Mater. **32**, 389–403 (2000)
46. Ramrakhyani, D.S., Lesieutre, G.A., Smith, E.C.: Modeling of elastomeric materials using nonlinear fractional derivative and continuously yielding friction elements. Int. J. Solids Struct. **41**, 3929–3948 (2004)
47. Naumenko, K., Altenbach, H.: Modeling of Creep for Structural Analysis. Springer, Berlin (2007)
48. Hyde, T.H., Sun, W., Wiliams, J.A.: Prediction of creep failure life of initially pressurized thick walled CrMoV pipes. Int. J. Press Vessels Pip. **76**, 925–933 (1999)
49. Staroselsky, A., Cassenti, B.N.: Combined rate-independent plasticity and creep model for single crystal. Mech Mater. **42**, 945–959 (2010)
50. Aktaa, J., Petersen, C.: Challenges in the constitutive modelling of the thermo-mechanical deformation and damage behaviour of EUROFER 97. Eng. Fract. Mech. **76**, 1474–1484 (2009)
51. Saleeb, A.F., Padula II, S.A., Kumar, A.: A multi-axial, multimechanism based constitutive model for the comprehensive representation of evolutionary response of SMAs under general thermomechanical loading conditions. Int. J. Plast. **27**, 655–687 (2011)
52. Auricchio, F., Taylor, R.L., Lubliner, J.: Shape-memory alloys: macromodelling and numerical simulations of the superelastic behaviour. Comput. Methods Appl. Mech. Eng. **146**, 281–312 (1997)
53. Dan, W.J., Zhang, W.G., Li, S.H., Lin, Z.Q.: A model for strain-induced martensitic transformation of TRIP steel with strain rate. Comput. Mater. Sci. **40**, 101–107 (2007)
54. Lee, M.G., Kim, S.J., Han, H.N., Jeong, WCh.: Implicit finite element formulations for multi-phase transformation in high carbon steel. Int. J. Plast. **25**, 1720–1758 (2009)

55. Lee, M.G., Kim, S.J., Han, H.N., Jeong, WCh.: Implicit finite element formulations for multiphase transformation in high carbon steel. Int. J. Plast. **25**, 1720–1758 (2009)
56. Field, J.E., Walley, S.M., Proud, W.G., Goldrein, H.T., Siviour, C.R.: Review of experimental techniques for high rate deformation and shock studies. Int. J. Impact. Eng. **30**, 725–775 (2004)
57. Das, A., Tarafder, S.: Experimental investigation on martensitic transformation and fracture morphologies of austenitic stainless steel. Int. J. Plast. **25**, 2222–2247 (2009)
58. Sedighi, M., Khandaei, M., Shokrollahi, H.: An approach in parametric identification of high strain rate constitutive model using Hopkinson pressure bar test results. Mater. Sci. Eng. A **527**, 3521–3528 (2010)
59. Crisfield, M.A.: Non-linear Finite Element Analysis of Solids and Structures. Advanced Topics, vol. 2. Wiley, Chichester (2001)
60. Bathe, K.J., Kojić, M.: Inelastic Analysis of Solids and Structures. Springer, Berlin (2005)
61. Wriggers, P.: Nonlinear Finite Element Methods. Springer, Berlin (2008)
62. Evans, L.C.: Partial Differential Equations, American Mathematical Society. Providence. Rhode Island (1998)
63. Rektorys, K.: Variační metody v inženýrských problémech a v problémech matematické fyziky. Academia, Praha (1999)
64. Bathe, K.J.: Finite Element Procedures in Engineering Analysis. Prentice-Hall Inc., Englewood Cliffs (1982)
65. Howell, J.R.: A catalog of radiation configuration factors. McGraw-Hill, New York (1976)
66. Écsi, L.: Numerical behaviour of a solid body under various mechanical loads using finite element method with new energy balance equation for fully coupled thermal-structural analysis. In: Proceedings of the Sixth International Congress on Thermal Stresses 2005, vol. 2, pp. 543–546. Technische Universität Wien, Wien (2005). ISBN 3-901167-12
67. Écsi, L, Élesztős, P.: Hysteretic heating of a solid bar using a universal constitutive equation with internal damping for fully coupled thermal-structural finite element analysis. In: Proceedings of the 8th International Congress Thermal Stresses 2009, vol. 1, pp. 233–236. University of Illinois Press, Illinois (2009). ISBN 978-0-615-28233-6
68. Écsi, L, Élesztős, P.: Trying to model the thermo-mechanical behaviour of a solid body under cyclic loading using a material model with internal damping. Acta Mechanica Slovaca Roč **12**(3-B), 143–150 (2008)
69. Écsi, L, Élesztős, P.: One of the possible variational formulations of fully coupled thermal-structural analysis. J. Mech. Eng. Roč **60**(3), 135–144 (2009)
70. Oden, T.J., Carey, F.G.: Finite Elements. A Second Course, vol. 2. Prentice-Hall Inc., Englewood Cliffs (1983)
71. Oden, T.J., Carey, F.G.: Finite Elements, Mathematical Aspects, vol. 4. Prentice-Hall Inc., Englewood Cliffs (1983)
72. Grutin, M.E.: The linear theory of elasticity. In: Flügge, S., Thruesdell, C. (eds.) Handbuch der Physic. VIa/2, pp. 1–295. Springer, New York (1972)
73. Oden, T.J.: Applied Functional Analysis. Prentice Hall, Englewood Cliffs (1979)
74. Reddy, B.D.: Introductory Functional Analysis. Springer, Berlin (1998)
75. Sun, C.T., Lu, J.P.: Vibration Damping of Structural Elements. Prentice Hall PTR, Englewood Clifs (1995)
76. Asszonyi, Csl, et al.: Izotrop kontinuumok anyagtulajdonságai. Műegyetemi kiadó, Budapest (2008)
77. Mesquita, A.D., Coda, H.B.: Alternative Kelvin viscoelastic procedure for finite elements. Appl. Math. Model. **26**, 501–516 (2002)
78. Argiris, J., Mlejnek, H.P.: Dynamics of Structures. Elsevier Science Publishers, Amsterdam (1991)
79. Bathe, K.J.: Finite Element Procedures. Prentice-Hall Inc., Englewood Cliffs (1995)
80. Lee, U.: Spectral Element Method in Structural Dynamics. John Wiley & Sons Pte Ltd., Singapore (2009)

81. Liu, M., Gorman, D.G.: Formulation of Rayleigh damping and its extensions. Comput. Struct. **57**(2), 277–285 (1995)
82. Belytschko, T., Liu, W.K., Moran, B.: Nonlinear finite elements for continua and structures. Wiley, Chichester (2000)
83. Écsi, L., Élesztős, P., Kosnáč, J.: Constitutive equation with internal damping for materials under cyclic and dynamic loadings using large strain/large deformation formulation. In: Proceedings of the International Conference on Computational Modelling and Advanced Simulations 2009, Bratislava, Slovak Republic (2009)
84. de Souza Neto, E.A., Perić, D., Owen, D.R.J.: Computational Methods for Plasticity. Theory and Applications. John Wiley & Sons Ltd., Singapore (2008)
85. Bonet, J., Wood, R.D.: Nonlinear Continuum Mechanics for Finite Element Analysis. Cambridge University Press, Cambridge (1997)
86. Écsi, L.: Extended NOIHKH model usage for cyclic plasticity of metals. Eng. Mech. Roč **13**(2), 83–92
87. Crisfield, M.A.: Non-linear Finite Element Analysis of Solids and Structures. Essentials, vol. 1. Wiley, Chichester (2000)
88. Trebuňa, F., Šimčák, F.: Príručka experimentálnej mechaniky. Edícia vedeckej a odbornej literatúry. TypoPress, Košice (2007)
89. Budó, Á.: Kísérleti fizika I. Nemzeti tankönyvkiadó, Budapest (1997)
90. Klepazko, J.R., Rusinek, A.: Experiments on heat generation during plastic deformation and stored energy for TRIP steels. Mater. Des. **30**, 35–48 (2009)
91. Johnson, W.A., Mehl, R.F.: Reaction kinetics in processes of nucleation and growth. Trans. AIME **135**, 416–458 (1939)
92. Avrami, M.: Kinetics of phase change I. J. Chem. Phys. **7**, 1103–1112 (1939)
93. Koistinen, D.P., Marburger, R.E.: A general equation prescribing the extent of the austenite-martensite transformation in pure iron-carbon alloys and carbon steels. Acta Metall. **7**, 59–60 (1959)
94. Lee, M.G., Kim, S.J., Han, H.N., Jeong, WCh.: Implicit finite element formulations for multi-phase transformation in high carbon steel. Int. J. Plast. **25**, 1726–1758 (2009)
95. Kang, S.H., Im, Y.T.: Three-dimensional thermo-elastic-plastic finite element modelling of quenching process of plain-carbon steel in couple with phase transformation. Int. J. Mech. Sci. **49**, 423–439 (2007)
96. Ronda, J., Oliver, G.J.: Consistent thermo-mechano-metallurgical model of welded steel with unified approach to derivation of phase evolution laws and transformation induced plasticity. Comput. Methods. Appl. Mech. Eng. **189**, 361–417 (2000)
97. Hömberg, D.: A numerical simulation of the Jominy end-quench test. Acta Matter. **44**(11), 4375–4385 (1996)

Performance of Hyperelastic Material Laws in Simulating Biaxial Deformation Response of Polypropylene and High Impact Polystyrene

K. Y. Tshai, E. M. A. Harkin-Jones and P. J. Martin

Abstract Free surface moulding processes such as thermoforming and blow moulding involve thermal and spatial varying rate dependent biaxial deformation of polymer. These processes are so rapid that the entire forming took place in a matter of seconds. As a result of the elevated rate of deformation, assumption that the deforming polymers experience no time dependent viscous dissipation or perfectly elastic up to large strain has became a common practice in numerical simulation. Following the above assumption, *Cauchy's* elastic and hyperelastic theories, originally developed for vulcanised natural rubber has been widely used to represent deforming polymeric materials in free surface moulding processes. To date, various methodologies were applied in the development of these theories, the most significant are those develop purely based on mathematical interpolation (mathematical models) and a more scientific network theories that involves the interpretation of macro-molecular structure within the polymer. In this chapter, the most frequently quoted *Cauchy's* elastic and hyperelastic theories, including Ogden, Mooney–Rivlin, neo-Hookean, 3-chain, 8-chain, Van der Waals full network, Ball's tube model, Edwards–Vilgis crosslinks-sliplinks model and the elastic model of Sweeney–Ward are reviewed. These models were analysed and fitted to a series of experimental high strain rate, high temperature, biaxial deformations data of polypropylene (PP) and high impact polystyrene (HIPS). The performance and

K. Y. Tshai (✉)
Department of Mechanical, Materials and Manufacturing Engineering, University of Nottingham Malaysia Campus, Jalan Broga, 43500 Semenyih, Selangor, Malaysia
e-mail: Kim-Yeow.Tshai@nottingham.edu.my

E. M. A. Harkin-Jones · P. J. Martin
School of Mechanical and Aerospace Engineering, Queen's University Belfast, Ashby Building, Stranmillis Road, Belfast BT9 5AH, Northern Ireland, UK
e-mail: e.harkinjones@qub.ac.uk

P. J. Martin
e-mail: p.j.martin@qub.ac.uk

N. Bonora and E. Brown (eds.), *Numerical Modeling of Materials Under Extreme Conditions*, Advanced Structured Materials 35, DOI: 10.1007/978-3-642-54258-9_9,

suitability of the various models in capturing the polymer's complex deformation behaviour during free surface moulding processes is presented.

Keywords Hyperelastic model · Biaxial deformation · High strain rate · High temperature · Free surface moulding

1 Introduction

Free surface moulding processes such as thermoforming and blow moulding are employed extensively in the food packaging industry, automobile applications, medical devices and most of our dairy appliances. It is widely recognised that this process is not utilised to its full potential. Optimisation of these processes are becoming technically challenging as newer process variations, materials, tighter sheet/preform and part tolerances, more critical applications and sophisticated controls are developed. At present, trial and error method is still widely practiced among small to medium manufacturing industries. Clearly a more scientific, efficient and sensible technique is required to reduce the excessive lead time and investment costs incurred during embodiment and detail design. To achieve these aims, the use of computer simulations to predict material behaviour under various processing conditions appears to be a very powerful tool. However, for these simulations to be useful it is important to integrate the analysis with an accurate material model. Utilising a high precision model simulating the material behaviour under actual forming conditions capable of effectively optimise the processing conditions as well as forming deployment, wall thickness, part shape and dimensional stability of the final part. In this chapter, the most frequently quoted *Cauchy's* elastic and hyperelastic theories, including Ogden, Mooney–Rivlin, neo-Hookean, 3-chain, 8-chain, Van der Waals full network, Ball's tube model, Edwards–Vilgis crosslinks-sliplinks model and the elastic model of Sweeney–Ward are reviewed. These models were analysed and fitted to a series of experimental high strain rate, high temperature, biaxial deformations data of polypropylene (PP) and high impact polystyrene (HIPS). The performance and suitability of the various models in capturing the polymer's complex deformation behaviour during free surface moulding processes is presented.

2 Experimental

2.1 Material

The thermoplastics tested in this work were pre-extruded HIPS sheet consisting white pigment (single batch production graded as Atofina ATO DPO 02) and nucleated, non-pigmented semi-crystalline isotactic homopolymer PP sheet with

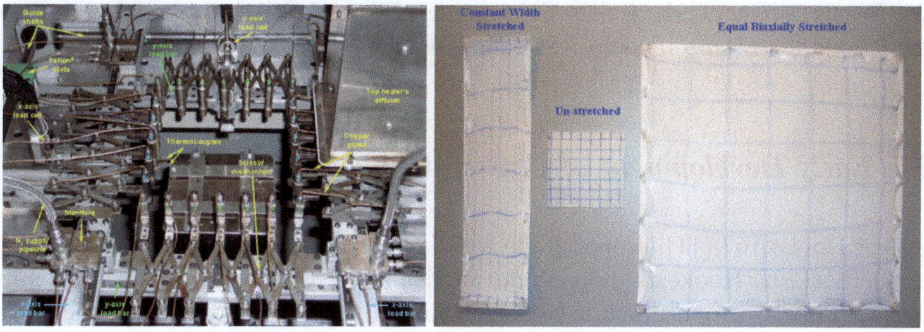

Fig. 1 Queen's biaxial stretcher (*left*); CW, un-stretched and EB stretched specimen (*right*)

anti-static agent (graded as Appryl 3030 BT1). The HIPS sheet has an average thickness of 1.51 ± 0.01 mm and MFI as quoted by the company was 4.5 (200 °C/5 kg, g/10 min) while the PP sheets has an average thickness of 1.42 ± 0.01 mm, MFI of 3 (230 °C/2.16 kg, g/10 min) and weight average molecular weight M_w as obtained by gel permeation chromatography was 339,000. The crystallinity and peak melting temperature of the PP sheet as characterised through DSC at heating rate of 10 °C/min were 59.6 % and 165 °C, respectively.

2.2 *Biaxial Test*

Equal biaxial (EB) and constant width (CW) deformation tests were carried out with the aids of a biaxial stretcher available at Queen's University Belfast. The pre-extruded specimens were tested under high nominal strain rates (ranging from 2 to 10 s^{-1}) and elevated temperatures (ranging from 145 to 160 °C for PP and 120 to 135 °C for HIPS). The un-stretched samples had a dimension 76 × 76 mm square and were pre-marked with 10 × 10 mm square ink grid to facilitate investigation of the local deformation on the post-stretched sample. Figure 1 shows the clamping frame of the biaxial stretcher and a comparison of the final shape obtained from CW and EB to that of an un-stretched sample.

3 Constitutive Modelling of Materials

In the history of material modelling, various constitutive equations were proposed in an attempt to represent the behaviour of particular polymer types depending on their morphological structure, i.e. amorphous or semi-crystalline. Within the literature, the available constitutive material models generally can be categorised into two major groups, hyperelastic and viscoelastic material laws. This chapter mainly concerned on the most frequently quoted *Cauchy's* elastic and hyperelastic theories, including Ogden, Mooney-Rivlin, neo-Hookean, 3-chain, 8-chain, Van der Waals full network, Ball's Tube model, Edwards-Vilgis crosslinks-sliplinks model and the elastic model of Sweeney-Ward.

3.1 Early Development of Classical Linear Elasticity

Robert Hooke 1678 [1], proposed the well-known Hooke's law of elasticity. For a perfectly elastic Hookean material, the applied force, F (hence stress) acting on its body is directly proportional to the extension, u (hence strain), with the inclusion of a material constant, k (this constant was later identified as the elastic or Young's modulus by Thomas Young in 1807 [2]), given as:

$$F = k \cdot u \tag{1}$$

French mathematician Augustin Cauchy 1789–1857 [3–6] subsequently generalised Hooke's law into three-dimensional elastic bodies and stated that the six components of stress are linearly related to the six components of strain, most commonly known as the constitutive Cauchy's elastic stress-strain relationship. The Cauchy's elastic stress-strain relationship written in the general form as:

$$\begin{bmatrix} \sigma_{11} \\ \sigma_{22} \\ \sigma_{33} \\ \sigma_{12} \\ \sigma_{13} \\ \sigma_{23} \end{bmatrix} = \begin{bmatrix} C_{1111} & C_{1122} & C_{1133} & C_{1112} & C_{1113} & C_{1123} \\ C_{2211} & C_{2222} & C_{2233} & C_{2212} & C_{2213} & C_{2223} \\ C_{3311} & C_{3322} & C_{3333} & C_{3312} & C_{3313} & C_{3323} \\ C_{1211} & C_{1222} & C_{1233} & C_{1212} & C_{1213} & C_{1223} \\ C_{1311} & C_{1322} & C_{1333} & C_{1312} & C_{1313} & C_{1323} \\ C_{2311} & C_{2322} & C_{2333} & C_{2312} & C_{2313} & C_{2323} \end{bmatrix} \begin{bmatrix} \varepsilon_{11} \\ \varepsilon_{22} \\ \varepsilon_{33} \\ \varepsilon_{12} \\ \varepsilon_{13} \\ \varepsilon_{23} \end{bmatrix} \tag{2}$$

where $[C_{ijkl}]$ is the constitutive stiffness matrix. There are 36 matrix components in general but conservative elastic materials possess a strain energy for a given strain state, as such, the constitutive stiffness matrix $[C_{ijkl}]$ is symmetric (i.e. $C = C^T$, only 21 independent matrix components in the generalised *Hooke's* law).

Taking the stress components as σ_{ij} and strain components as ε_{ij}, if three planes of symmetry exist and if the coordinate planes are parallel to these planes, the Cauchy's elastic stress-strain relationship can be organised into a simpler form, given as:

$$\begin{bmatrix} \sigma_{11} \\ \sigma_{22} \\ \sigma_{33} \\ \sigma_{12} \\ \sigma_{13} \\ \sigma_{23} \end{bmatrix} = \begin{bmatrix} C_{1111} & C_{1122} & C_{1133} & 0 & 0 & 0 \\ C_{1122} & C_{2222} & C_{2233} & 0 & 0 & 0 \\ C_{1133} & C_{2233} & C_{3333} & 0 & 0 & 0 \\ 0 & 0 & 0 & C_{1212} & 0 & 0 \\ 0 & 0 & 0 & 0 & C_{1313} & 0 \\ 0 & 0 & 0 & 0 & 0 & C_{2323} \end{bmatrix} \cdot \begin{bmatrix} \varepsilon_{11} \\ \varepsilon_{22} \\ \varepsilon_{33} \\ \varepsilon_{12} \\ \varepsilon_{13} \\ \varepsilon_{23} \end{bmatrix} \tag{3}$$

where the stiffness coefficient C_{ijkl} is a fourth order tensor (denoting the ratio of stress to strain) with only 9 independent coefficients and the material is referred as orthotropic symmetric.

While the generalised Hooke's law is capable of describing deformation of a material within its linear elastic region, this model failed to simulate the non-linear, large deformation behaviour of polymeric materials. To overcome this limitation, various forms of constitutive equations have been proposed and this has led to the development of a number of non-linear elastic or hyperelastic models. The fundamental concept of these formulations is generally based on the conservation of energy. The theory states that any work done (in a reversible isothermal process) in the deformation of a material is stored as potential energy (frequently called the free energy potential or strain energy) within the deforming body, and a deformed material will completely recover to its initial un-stretched state on removal of the forming force. In other words, the nonlinear strain energy for a given strain state only depends on the strain state itself and not on the manner in which this strain state was obtained. The many forms of nonlinear strain energy function W generally can be categorised into two major groups depending on the fundamental concepts employed in their development, namely phenomenological and physical network hyperelastic material laws.

3.2 Phenomenological Hyperelastic Material Laws

3.2.1 Mooney-Rivlin Model

Within the many forms of phenomenological strain energy functions, the earliest and simplest is the *Mooney* form. Assuming a material is incompressible and isotropic, Mooney 1940 [7] derived by purely mathematical arguments, a representation of the strain energy, W.

$$W = C_1\left(\lambda_1^2 + \lambda_2^2 + \lambda_3^2 - 3\right) + C_2\left(\lambda_1^{-2} + \lambda_2^{-2} + \lambda_3^{-2} - 3\right) \tag{4}$$

and

$$\lambda_i = \frac{L_i + \Delta L_i}{L_i} \tag{5}$$

where λ_i (i = 1,2,3) are the principal stretch ratios given as the ratio of current length $(L + \Delta L)$ over the original length L in the three principal directions respectively, C_1 and C_2 are two elastic constants.

Once the strain energy function W has been defined, assuming that the material is incompressible, the principal true stresses σ_T can be computed (by differentiating W with respect to each of the three principal stretch ratios, i = 1,2,3) from W to within an arbitrary pressure, P, where P can usually be determined from the boundary conditions (in order to eliminate the pressure term, the true stress-stretch ratio relations are frequently written in terms of the difference in two principal stresses, i.e. $\sigma_1 - \sigma_2$, $\sigma_2 - \sigma_3$ and $\sigma_3 - \sigma_1$):

$$(\sigma_T)_i = \lambda_i \cdot \frac{\partial W}{\partial \lambda_i} + P \tag{6}$$

Eight years later in 1948, Rivlin [8, 9] claimed that the strain energy function W should be formulated such that it is symmetrical with respect to the three principal stretch ratios for the principle of material objectivity to be satisfied. In this context, the author proposed that the strain energy function must depend only on the even powers of the three stretch ratios. This has led to the development of three strain invariants (I_1, I_2, and I_3) as even power functions of λ_i ($i = 1, 2, 3$), representing strain fields that are independent on the particular choice of coordinate system.

$$\begin{aligned} I_1 &= \lambda_1^2 + \lambda_2^2 + \lambda_3^2 \\ I_2 &= \lambda_1^2\lambda_2^2 + \lambda_2^2\lambda_3^2 + \lambda_3^2\lambda_1^2 \\ I_3 &= \lambda_1^2\lambda_2^2\lambda_3^2. \end{aligned} \tag{7}$$

For an incompressible material, the third strain invariant $I_3 = 1$ since $\lambda_1\lambda_2\lambda_3 = 1$, leaving only two independent strain invariants (I_1 and I_2) in the system. The author subsequently proposed the use of only first and second invariants of strain in the strain energy function and stated that this function could be expanded as an infinite series as show below.

$$W = \sum_{i=0,j=0}^{\infty} C_{ij}(I_1 - 3)^i (I_2 - 3)^j \tag{8}$$

where C_{ij} are material constants with C_{00} being equal to 0.

However, when considering only the first term, the *Rivlin* strain energy function for an incompressible material (Eq. 8) the *Mooney* form is obtained. This is generally known as the first order *Mooney-Rivlin* or simply the *Mooney-Rivlin* relationship for higher order terms.

$$W = C_{10}(I_1 - 3) + C_{01}(I_2 - 3) \tag{9}$$

3.2.2 Ogden Model

In a slightly later development, Ogden 1972 [10] employed a more formal treatment by formulating the strain energy function directly in terms of the three principal stretch ratios. The author proposed that the 'power (constant α_i in Eq. 10

below)' applied to the three principal stretch ratios may have any values, positive or negative and are not necessarily integers, given in the form below.

$$W = \sum_{i=1}^{N} \frac{2\mu_i}{\alpha_i^2} \left(\lambda_1^{\alpha_i} + \lambda_2^{\alpha_i} + \lambda_3^{\alpha_i} - 3 \right) \tag{10}$$

where μ_i and α_i are material constants determined from the experimental data and $i = 1, 2, 3, \ldots, N$, representing the order of the strain energy function.

It should be noted that under unique case where $N = 1$ and constant $\alpha_i = 2$, the *Ogden* model yields a special case of the *Mooney* form with constant $C_2 = 0$.

3.3 Physical Network Hyperelastic Material Laws

In physical network hyperelastic material laws, polymers are considered as consisting of numerous long flexible chains, each of which is capable of assuming a variety of configurations in response to the thermal vibrations of '*micro-Brownian*' motion of their constituent atoms. Furthermore it is assumed that the molecular chains are interlinked so as to form a coherent network but that the number of cross-links is relatively small and is not sufficient to significantly interfere with the motion of the chains. In the unstrained state, these chain molecules always tend to assume a set of more stable configurations corresponding to a state of maximum entropy. In the case where the chain molecules are constrained by external forces, their configuration will be changed to produce a state of strain. The probability of potential stable chain configurations in the unstrained state has been widely studied using statistical mechanics and assuming that deformation processes at the strained state are always thermodynamically reversible [11, 12].

3.3.1 Gaussian Network Model

neo-Hookean Model

In 1943, Treloar [13, 14] with the assumption that material being incompressible and isotropic, developed the *neo-Hookean* strain energy function by considering the macroscopic molecular configuration of rubber-like materials with *Gaussian* chain length distribution. Since *Gaussian* statistical analysis was the earliest concept used to develop physically based models and this concept is closely related to the development of other physical models, describing this theorem in further detail will provide a better overall picture and aid in the understanding of later developments in this area.

The *Gaussian* network assumes that:

1. The molecular network within a material contains N chains per unit volume, a chain being defined as the segment of a molecule between successive points of cross-linkage.
2. Within a particular chain, there are n 'freely jointed' carbon-carbon (C–C) links/segments, each of length l. The terms freely jointed chain implies that a chain consists of equal C–C links joined together without the restriction that the valence angles should remain constant, a case where random joining or 'random walk' is assumed.
3. The mean-square end-to-end distance for the whole assembly of chains in the unstrained state is the same as for a corresponding set of free chains (end-to-end distance r being very much smaller compared to the fully extended length of the chain).
4. The junction points between chains move on deformation as if they were embedded in an elastic continuum. As a result, the components of length of each chain change in the same ratio as the corresponding dimensions of the bulk rubber, known as 'affine' deformation assumption.
5. The entropy of the network is the sum of the entropies of all individual chains.

Based on the assumptions of a *Gaussian* network, Treloar statistically calculated for a single molecular chain, the number of possible configurations corresponding to a chosen end-to-end distance described in assumption 3 above as well as its entropy. The change in entropy of the whole network was successively computed by the entropy summation of N chains as described in assumption 5. Figure 2 schematically shows a single molecular chain with $n = 36$ freely jointed segments/links of length l, where the probability of the chain assuming end-to-end distance r, according to assumption 3, was statistically evaluated on the basis of spherically symmetrical distribution ($x^2 + y^2 + z^2 = r^2$).

Under a *Gaussian* statistical treatment, the probability that chain end B will fall within an elemental volume between r and dr of a spherical shell of radius r and thickness dr was obtained.

$$P(r) \cdot dr = \frac{b^3}{\pi^{\frac{3}{2}}} \exp\left(-b^2 r^2\right) \cdot 4\pi r^2 \cdot dr \tag{11}$$

where b is a function of both the number of segments/links n and their average length l, given as $b^2 = [3/(2\,nl^2)]$.

By evaluating the most probable value from Eq. 11 (differentiating $p(r)$ with respect to r), it was subsequently found that the mean-square value of r, $(r_{mean})^2$, is $n \cdot l^2$ and thus the un-deformed root-mean-square end-to-end distance, $\left[\sqrt{(r_{mean})^2}\right]$, of any arbitrary molecular chain can be given as $\sqrt{n} \cdot l$. Furthermore it is clear from both *Gaussian* network assumptions and Fig. 2 that the fully stretched length of a freely jointed chain can be given by $n \cdot l$. According to

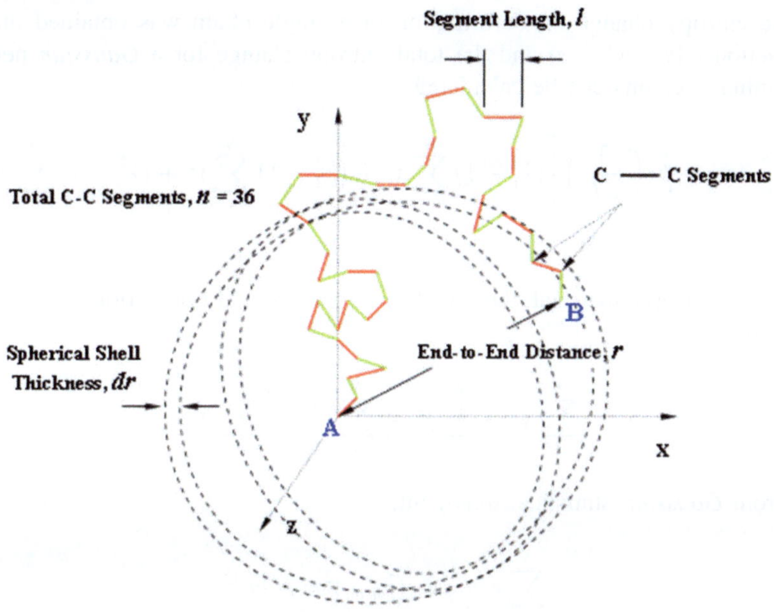

Fig. 2 Single molecular chain with 36 freely jointed segments

Boltzmann's statistical thermodynamics, the entropy of such a freely jointed chain is proportional to the natural logarithm of the number of possible configurations, which in turn can be given in terms of unit volume probability or probability density (e.g. normalised with respect to the spherical shell volume '$4\pi r^2$ '), *p(r)*,

$$s = k_b[\ln p(r)] \tag{12}$$

where k_b is the *Boltzmann's* constant equal to 1.38×10^{-23} JK^{-1}.

Substituting *p(r)* from Eq. 11 into Eq. 12, a function for the entropy of an undeformed freely jointed chain can be expressed as:

$$s = k_b\left[\ln(c) - b^2r^2\right] = c - k_b \cdot \left(\frac{3}{2nl^2}\right) \cdot \left(x^2 + y^2 + z^2\right) \tag{13}$$

where *c* is an arbitrary constant.

On deformation, where the three principal stretch ratios λ_1, λ_2 and λ_3 were chosen to be parallel to the three rectangular axes *x*, *y* and *z*, the entropy of the chain changes to *s'*.

$$s' = c - k_b \cdot \left(\frac{3}{2nl^2}\right) \cdot \left(\lambda_1^2x^2 + \lambda_2^2y^2 + \lambda_3^2z^2\right) \tag{14}$$

The entropy change on deformation of a single chain was obtained through subtraction ($\Delta s = s' - s$) and the total entropy change for a *Gaussian* network containing N chains can be calculated as:

$$\sum_{1}^{N} \Delta s = -k_b\left(\frac{3}{2nl^2}\right)\left[\left(\lambda_1^2 - 1\right)\sum_{1}^{N} x^2 + \left(\lambda_2^2 - 1\right)\sum_{1}^{N} y^2 + \left(\lambda_3^2 - 1\right)\sum_{1}^{N} z^2\right] \tag{15}$$

For an isotropic material, there will be no preferential orientation for the x, y or z directions.

$$\sum_{1}^{N} x^2 = \sum_{1}^{N} y^2 = \sum_{1}^{N} z^2 = \frac{1}{3}\sum_{1}^{N} r^2 \tag{16}$$

and from *Gaussian* statistical treatment,

$$\sum_{1}^{N} r^2 = N \cdot (r_{mean})^2 = N \cdot nl^2 \tag{17}$$

Assuming that the internal energy remains constant on deformation, substitution of Eqs. 16 and 17 into 15 produces the *Helmholtz* free energy for an incompressible, isotropic material under reversible deformation, in terms of the *neo-Hookean* form below.

$$W = -T \cdot \Delta s = \frac{1}{2} N k_b T\left(\lambda_1^2 + \lambda_2^2 + \lambda_3^2 - 3\right) \tag{18}$$

where $N \cdot k_b \cdot T$ is the material's shear modulus G and T is the absolute temperature.

It should be noted that the strain energy function derived by Treloar (Eq. 18) yields a form of the *Mooney* strain energy function when constant C_2 in Eq. 4 equal to 0.

When fitted the *neo-Hookean* model to experimental data [11], reasonably good agreement was obtained in the low strain region but the model's accuracy gradually decreased with increasing strain. This problem was attributed to the possibility that *Gaussian* distribution function may be valid only as long as the end-to-end distance of the chain is not so large as to be comparable with its fully extended length (generally end-to-end distance r should not be more than 1/3 of the fully extended chain length).

3.3.2 Non-Gausian Network Models

At about the same time as the *neo-Hookean* model was proposed, Kuhn and Grun [15] attempted to describe the probability of the configuration of molecular chains based on a *non-Gaussian* statistical theory. In this study, many assumptions of the *Gaussian* network remained (assumptions 1, 2, 4 and 5) with the exception of the assumption 3 which required the end-to-end distance to be very much less than the fully extended chain length was removed. This was done by replacing the *Gaussian* statistical distribution with an 'inverse *Langevin*' approximation function for the computation of probability distribution. Under inverse *Langevin* statistical analysis, Kuhn and Grun demonstrated that the probability density *p(r)* for chain end-to-end distance r could be evaluated as:

$$\ln p(r) = c - n\left(\frac{r}{nl}\beta + \ln\frac{\beta}{\sinh\beta}\right) \tag{19}$$

where c is an arbitrary constant and β is a function determined by the fractional extension of the chain *(r/nl)*, given in the form below.

$$\frac{r}{nl} = \coth\beta - \frac{1}{\beta} = \boldsymbol{L}(\beta) \tag{20}$$

The function L is the *Langevin* function and parameter β can be obtained from the inverse of the *Langevin* function L^{-1}, which in turn can be expressed in a series form as shown below:

$$\beta = \boldsymbol{L}^{-1}\left(\frac{r}{nl}\right) = \boldsymbol{L}^{-1}\left(\coth\beta - \frac{1}{\beta}\right) \tag{21}$$

$$\Rightarrow \beta = 3\left(\frac{r}{nl}\right) + \frac{9}{5}\left(\frac{r}{nl}\right)^3 + \frac{297}{175}\left(\frac{r}{nl}\right)^5 + \frac{1539}{875}\left(\frac{r}{nl}\right)^7 + \cdots \tag{22}$$

By substitution of β from Eq. 22 into Eq. 19, the authors obtained a definition for the density *p(r)*, which in turn can be expressed as an expansion series.

$$\ln p(r) = c - n\left[\frac{3}{2}\left(\frac{r}{nl}\right)^2 + \frac{9}{20}\left(\frac{r}{nl}\right)^4 + \frac{99}{350}\left(\frac{r}{nl}\right)^6 + \cdots\right] \tag{23}$$

Adopting an analysis identical to *Boltzmann's* statistical thermodynamics, the entropy of a stretched chain with current end-to-end distance r was obtained as:

$$s = k_b \ln p(r) = c - n \cdot k_b\left[\frac{3}{2}\left(\frac{r}{nl}\right)^2 + \frac{9}{20}\left(\frac{r}{nl}\right)^4 + \frac{99}{350}\left(\frac{r}{nl}\right)^6 + \cdots\right] \tag{24}$$

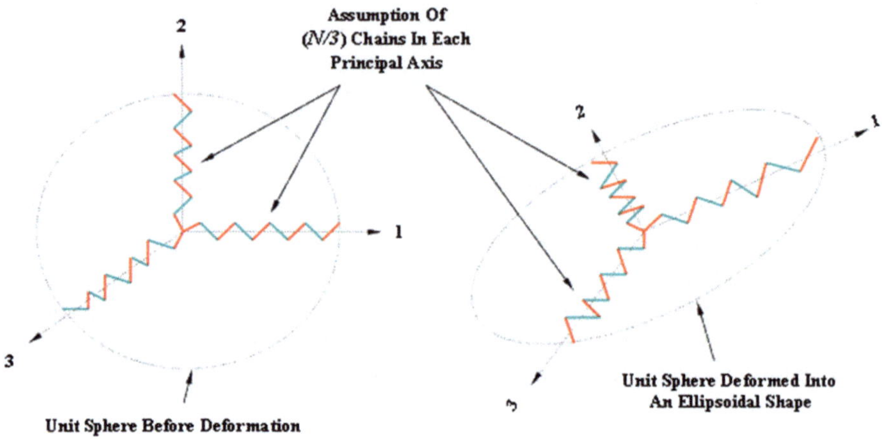

Fig. 3 Network configuration of the three-chain model: noting that the three polymer chains are always oriented in the principal strain axes before and during deformation

For a molecular network containing *N* chains per unit volume, the total tension of the network can be evaluated as the summation of the differentiated entropy function (with respect to *r*).

$$f = \sum_{1}^{N} -T\left(\frac{\partial s}{\partial r}\right) = \frac{Nk_bT}{l}\left[3\left(\frac{r}{nl}\right) + \frac{9}{5}\left(\frac{r}{nl}\right)^3 + \frac{297}{175}\left(\frac{r}{nl}\right)^5\right] = \frac{Nk_bT}{l}\boldsymbol{L}^{-1}\left(\frac{r}{nl}\right) \tag{25}$$

It should be noted that the *Gaussian* formula in Eq. 13 is similar to the first term of Eq. 24 above. This suggested that *Gaussian* statistical theory assumed all higher terms in *(r/nl)* negligible compared to the first term. In the case where the ratio *(r/nl)* increases with increasing chain extension, the prediction from the *Gaussian* distribution function gradually diverges from the 'inverse *Langevin*' statistical treatment, leading to the failure of the *Gaussian* formula at large strain.

Three-Chain Model

The superiority of the *non-Gaussian* over the *Gaussian* statistical theory gave rise to the later development of a network model based on this treatment. Guth et al. [16, 17] who employed a similar concept, developed the three-chain model based on assumption that in a molecular network containing *N* chains per unit volume *(N/3)* polymer chains per unit volume are always oriented in each of the three principal strain axes such that the system can be represented by 3 identical chains, as shown in Fig. 3.

Therefore, the total network strain energy can be expressed as the sum of the 3 single-chain strain energy functions weighted by the factor *(N/3)*. The three-chain model defined in terms of the principal *Cauchy* stress and stretch ratios relationship can be given as:

$$\sigma_i^{3-chain} = \frac{1}{3}(Nk_bT)\sqrt{n} \cdot \lambda_i \cdot \boldsymbol{L}^{-1}\left(\frac{\lambda_i}{\sqrt{n}}\right) + P \tag{26}$$

where $(N \cdot k_b \cdot T)$ according to *Gaussian* statistics is a material parameter defining its shear modulus, parameter $(\sqrt{n})$ is a constant obtained from inverse *Langevin* treatment as the ratio of fully extended chain length $(n \cdot l)$ over its initial unstretched length $(\sqrt{n}) \cdot l$, denoted as the 'limiting stretch or finite inextensibility' of the chain λ_{max}. This corresponds to an infinite stress in terms of the model.

Eight-Chain Model

In order to facilitate finite element coding, Arruda and Boyce 1993 [18] developed a simplified form of the full-network model [19–21] where a constitutive relation without the need of numerical integration was developed based on an eight chain representation of the underlying macromolecular network and each individual chain property was described through the use of *non-Gaussian* inverse *Langevin* theory. The proposed model assumed that in a molecular network containing an assembly of N chains per unit volume, each chain consists of n links of length l, there are *(N/8)* chains per unit volume being arranged in such a way that the system can be represented by 8 identical chains linked at the centre of an enclosed symmetric cube in principal space and their outer ends being fixed at the eight corners of the cube. On deformation, the cube is always oriented in the principal frame and the length of each enclosed chain was evaluated from the three principal stretch ratios λ_1, λ_2, and λ_3 of the cube edges, as shows in Fig. 4.

The eight-chain model defined in terms of the principal *Cauchy* stress and stretch ratios relationship as

$$\sigma_i^{8-chain} = \frac{1}{3}(Nk_bT)\sqrt{n}\frac{\lambda_i^2}{\lambda_{chain}}\boldsymbol{L}^{-1}\left(\frac{\lambda_{chain}}{\sqrt{n}}\right) + P \tag{27}$$

where material constants (Nk_bT) and $(\sqrt{n})$ as well as the subscript i remain as defined in Eq. 26. The term λ_{chain} is the extension ratio (r/r_0) of the eight chains, defined in terms of the three principal stretch ratios as:

$$\lambda_{chain} = \sqrt{\frac{\lambda_1^2 + \lambda_2^2 + \lambda_3^2}{3}} \tag{28}$$

where λ_1, λ_2, and λ_3 are the principal stretch ratios of the cube edges.

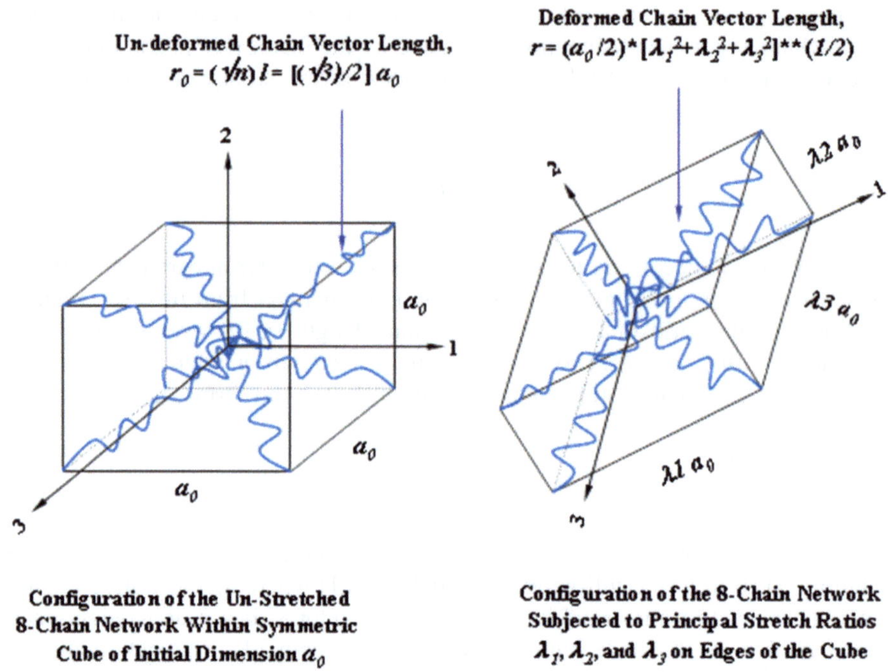

Fig. 4 Schematic diagram of 8-chain model, un-stretched and stretched

Ball "Tube" Model

In agreement with the opinions of Flory and Erman [22], Ball, Edwards and co-workers [23, 24] claimed that the classical theories of rubber elastic networks [16] were unrealistic as they assumes the network chains to be held only by numerous cross-linkages and these molecular chains are capable of moving through each other as 'phantoms' under the action of external applied forces. However, in a real polymeric structure, these molecular chains are capable of rotation, bending, and kinking about their chemical back-bone according to their steric constraints, in which a large assembly of randomly coiled chains may forms an enormous entanglement in addition to the cross-linkages afore mentioned. This topological entanglement gives an additional contribution to the elasticity of rubber and therefore must be completely preserved in the evaluation for a more physical strain energy function. The proposed model assumed that in a dense network consisting of high molecular weight, the motion of two entangled neighbouring chains (each of which has a fully extended chain length between cross-linkages of L) are essentially confined in a tube-like region called slip-links, made of the large number of surrounding molecular chains. A schematic representation of the proposed 'tube model' can be visualised in Fig. 5.

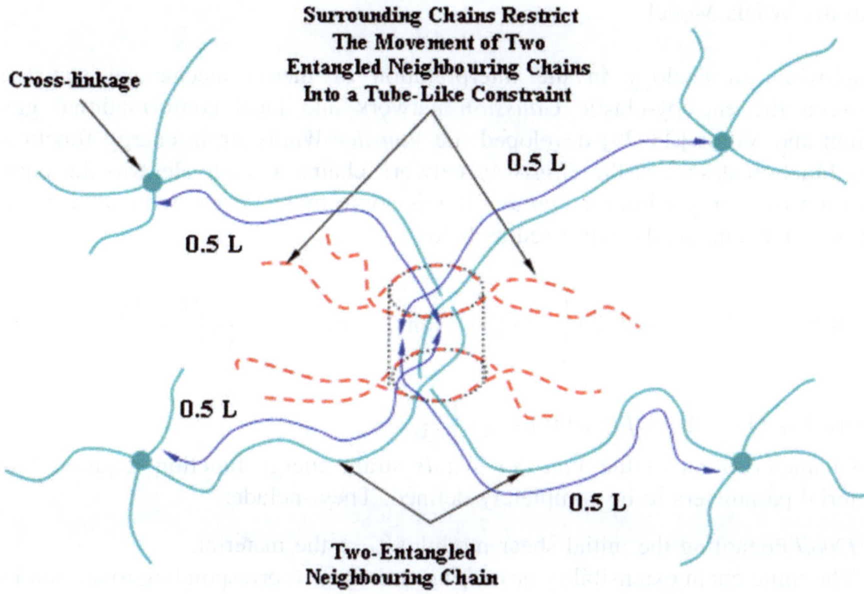

Fig. 5 Schematic representation of 'tube model' consisting of both chain entanglement and cross-linkages as proposed by Ball, Edwards and co-workers

It is clear from Fig. 5 that under the concept of an effective tube of constraint, the sliding freedom of any entangled chains is thus only an arc length 0.5 L in any one direction until it locks onto another entanglement or cross-linkage. Through a rather lengthy treatment of '*replica formalism*', the authors attained a definition for the strain energy, W, of the '*tube* model', involving both the contributions from enormous cross-linkages and chain entanglements. The resulting change in strain energy (per unit volume) can be obtained by integrating the differentiated form ($dW/d\lambda$) of Eq. 29 below.

$$W = \frac{1}{2}(N_c k_b T)\sum_{i=1}^{3}\lambda_i^2 + \frac{1}{2}(N_s k_b T)\sum_{i=1}^{3}\left[\frac{(1+\eta)\lambda_i^2}{1+\eta\lambda_i^2} + \ln\left(1+\eta\lambda_i^2\right)\right] \tag{29}$$

where parameters N_c and N_s are related to the number of crosslinks and sliplinks per unit volume respectively, k_b is the *Boltzmann's* constant equal to 1.38×10^{-23} JK^{-1}, T is the absolute temperature, η is a measure used to define the freedom to slide for the sliplinks and its value is either greater or equal to 0. In the extreme cases where $\eta = 0$ (equivalent to zero slippage and value L in Fig. 5 equal to 0), the change in strain energy reduces to a state depending only on the contribution from cross-linkages alone.

Van der Waals Model

Employing an analogy in the interpretation of thermo-mechanical statistics between the entropy-elastic *Gaussian* network and ideal conformational gas, Kilian and Vilgis [25–29] developed the *Van der Waals* strain energy function. The function describes the *Gaussian* network chains as equivalent to the equipartition of energy within a *Van der Waals* conformational gas with weak interactions, mathematically expressed as below.

$$W^{Van-der-Waals} = (Nk_bT)\left\{-(\lambda_m^2 - 3)[\ln(1-\eta)+\eta] - \frac{2}{3}a\left(\frac{\tilde{I}-3}{2}\right)^{\frac{3}{2}}\right\} \quad (30)$$

where $\tilde{I} = (1+\beta)I_1 + \beta I_2$ and $\eta = \sqrt{\frac{\tilde{I}-3}{\lambda_m^2-3}}$

Characterisation of the *Van der Waals* strain energy function required four material parameters to be completely defined. They include:

1. *(Nk_bT)* denoting the initial shear modulus G of the material.
2. The finite chain extensibility or locking stretch λ_m (corresponding to an infinite strain energy potential when its limiting value is reached, or more precisely when $\tilde{I} \rightarrow {\lambda_m}^2$).
3. Parameter a characterising the global interaction between 'quasi-particle' of an ideal gas, or the global interaction between chains in the interpretation of a rubber elastic network.
4. Dimensionless constant β represents a linear mixture parameter combining strain invariants I_1 and I_2 into $\tilde{I}$. It should be noted that when $\beta = 0$, the *Van der Waals* potential will depends on the first invariant alone.

Edwards-Vilgis Crosslink-Sliplink Model

Edwards and Vilgis [30, 31] attempted to improve the previously proposed 'tube model' through a detailed treatment of the consequences of entanglements. The authors found that apart from the replica calculation employed by Ball et. al. [23], the strain energy function for a network consisting of both cross-links and slip-links similar to that of 'tube model' can be evaluated through several other methods including (1) *Flory* segment argument, (2) inextensibility limit of a single chain that interpret a chain's maximum locking stretch as the difference between its fully extended length L and the 'primitive path' length of the tube L_{pp} and/or (3) the *Rouse* theory of linear viscoelasticity [32] (the readers are referred to the literatures [30, 31, 33, 34] for further details in each of these concepts). The resulting unit volume strain energy function involves an additional parameter α, which measure the inextensibility λ_{max} (corresponding to a singularity in stress) of the network chains *($\alpha = 1/\lambda_{max}$)*, expressed in the form given below

$$W^{Edwards-Vilgis} = W_{CL} + W_{SL} \tag{31}$$

where the strain energy contribution from cross-links W_{CL} is

$$W_{CL} = \frac{1}{2}(N_c k_B T)\left[\frac{\sum_{i=1}^{3}(1-\alpha^2)\lambda_i^2}{1-\alpha^2\sum_{i=1}^{3}\lambda_i^2} + \ln\left(1-\alpha^2\sum_{i=1}^{3}\lambda_i^2\right)\right] \tag{32}$$

and the strain energy contribution as a result of entanglement or slip-links is

$$W_{SL} = \frac{1}{2}(N_s k_B T)\left\{\sum_{i=1}^{3}\left[\frac{(1+\eta)(1-\alpha^2)\lambda_i^2}{\left(1-\alpha^2\sum_{i=1}^{3}\lambda_i^2\right)(1+\eta\lambda_i^2)} + \ln(1+\eta\lambda_i^2)\right] + \ln\left(1-\alpha^2\sum_{i=1}^{3}\lambda_i^2\right)\right\} \tag{33}$$

The material's constants N_c, N_s, and η are as defined in Eq. 29. It should be noted that under the extreme case where $\eta = 0$, the energy contribution from slip-links (Eq. 33) returns a form similar to that of the cross-links contribution (Eq. 32). Under such an extreme case, there is no chain slippage allowed in the entire network and the entanglements act as permanent cross-links.

Sweeney-Ward Model

In the work of Sweeney and Ward [35–37], the authors observed that a necking phenomenon routinely occurred in the drawing of semi-crystalline PP and subsequently proposed the modelling of such instabilities by exploiting a modified version of the '*tube*' model. In a uniaxial stretching simulation of PP incorporating the 'tube' theory, it was found that the model successfully captured the many features of the empirical data. However, the model could not reproduce the onset of necking at extension ratios λ less than ≈ 1.8; in reality empirical results had shown that the necking of PP initiated at λ as early as ≈ 1.3. In order to allow the modelling of the onset of necking at a strain comparable to those observed experimentally, a modification was proposed such that as deformation proceeds, interaction between sliplinks in the immediate neighbourhood may lead to a decrease in the sliplinks number N_s. This has been done by making the previously constant sliplink number N_s dependent on the first invariant of strain I_1

$$N_s = \frac{N_{s0} - N_{sf}}{(I_1 - 2)^{\beta}} + N_{sf} \tag{34}$$

where N_{s0} and N_{sf} are constants corresponding to the initial and ultimate sliplink numbers respectively ($N_{sf} \leq N_s \leq N_{s0}$), and β is simply an additional fitting parameter that controls the rate of decay of N_s.

Assuming incompressibility and expressing the three principal stretch ratios λ_1, λ_2, λ_3 in terms of the three invariants of strain I_1, I_2 and I_3, the strain energy function of the '*tube*' theory can be rewritten as

$$\frac{W}{k_B T} = \frac{N_c}{2} I_1 + \frac{N_s}{2}\left[\frac{(1+\eta)(I_1 + 2\eta I_2 + 3\eta^2)}{1 + \eta I_1 + \eta^2 I_2 + \eta^3} + \ln\left(1 + \eta I_1 + \eta^2 I_2 + \eta^3\right)\right] \quad (35)$$

It should be noted that in the evaluation of stress, the expressions ($\partial W/\partial I_1$) and ($\partial W/\partial I_2$) were first obtained by treating N_s as constant. The varying N_s function (as shown in Eq. 34) entered into the proposed model through the differentiated functions of ($\partial W/\partial I_1$) and ($\partial W/\partial I_2$).

3.4 Characterisation of Material Models

The behaviour of a particular polymeric system might well be described by a specific model but not the others. The effectiveness of the chosen material models were assessed in terms of their accuracy in predicting the experimental material behaviour of the PP and HIPS used. The characterisations were carried out through a nonlinear least square fit procedure where the sum of square of the error measure, E, is to be minimised.

$$E = \sum_{i=1}^{N} \left(\sigma_i^{test} - \sigma_i^{th}\right)^2 \quad (36)$$

where N is the number of experimental true stress-stretch ratio data pairs, σ_i^{test} is true stress value from experimental data and σ_i^{th} is model generated true stress.

A number of assumptions were adopted in the evaluation process:

1. The material is assumed to be fully incompressible, i.e. the volume of the material cannot change throughout the deformation process ($\lambda_1 \cdot \lambda_2 \cdot \lambda_3 = 1$).
2. The material is assumed to be homogeneous and isotropic.
3. The material is assumed to be perfectly elastic up to large strain.
4. All nonlinear least square fitting were performed on experimental CW and EB deformation data of PP (at 150 °C, 2 s^{-1}) and HIPS (at 130 °C, 2 s^{-1}).

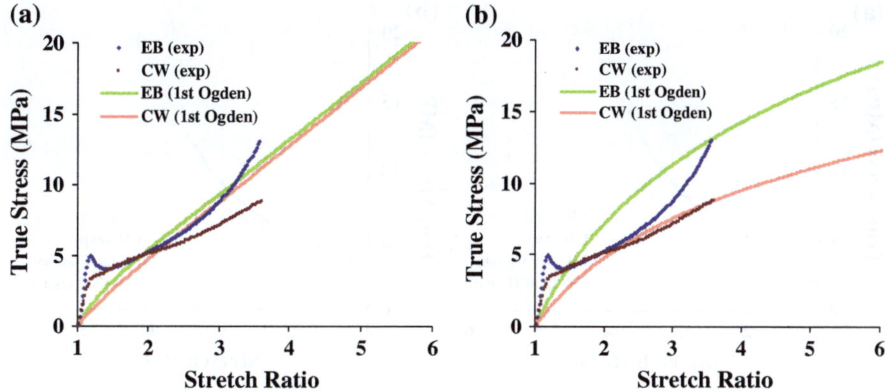

Fig. 6 Model prediction (1st order *Ogden*) to EB and CW responses of PP. **a** Fitted to equal biaxial test data. **b** Fitted to constant width test data

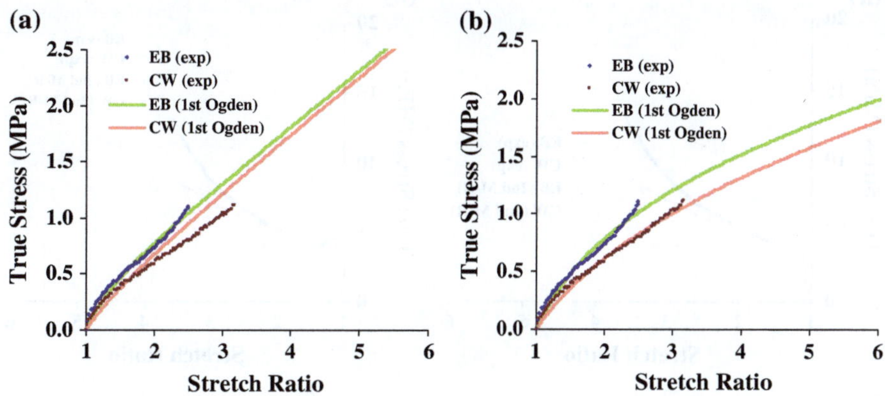

Fig. 7 Model prediction (1st order *Ogden*) to EB and CW responses of HIPS. **a** Fitted to equal biaxial test data. **b** Fitted to constant width test data

4 Results and Discussions

4.1 Phenomenological Hyperelastic Model

4.1.1 Ogden Model

The 1st order *Ogden* model was least-square fitted to both EB and CW data of PP and HIPS. Figures 6 and 7 show the fitted results of the 1st order Ogden model.

While the 1st order Ogden model predict relatively well the deformation behaviour of HIPS, the model failed to capture the deformation response of the PP. It can be observed that a higher order terms (2nd order Ogden model) introduced

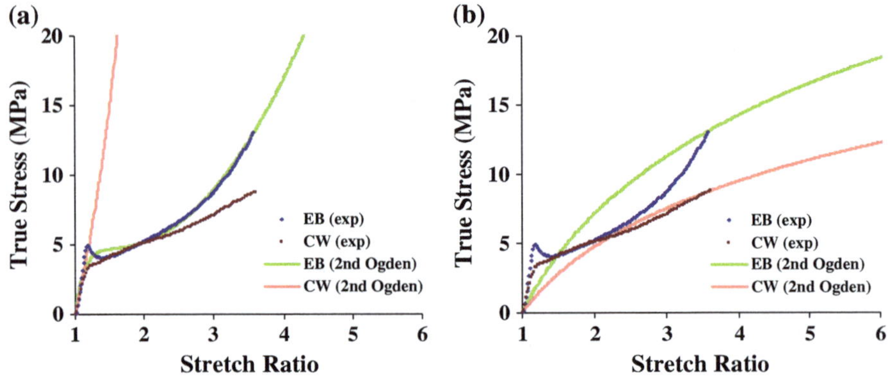

Fig. 8 Model prediction (2nd order *Ogden*) to EB and CW responses of PP. **a** Fitted to equal biaxial test data. **b** Fitted to constant width test data

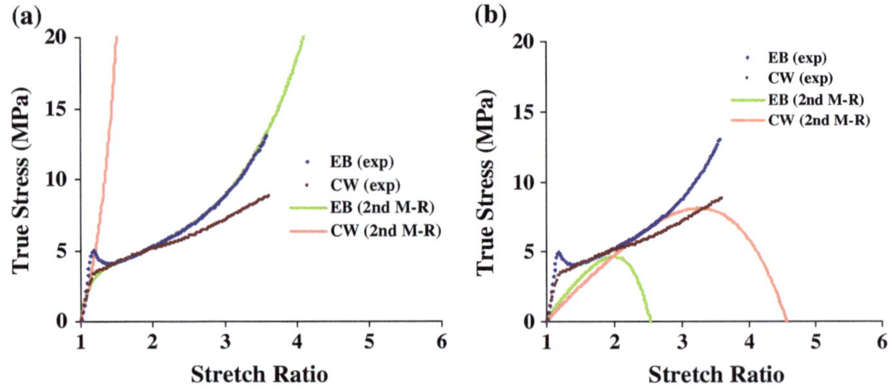

Fig. 9 Model prediction (2nd order M-R) to EB and CW responses of PP. **a** Fitted to equal biaxial test data. **b** Fitted to constant width test data

greater nonlinearity into the model simulated true stress-stretch ratio curves, Fig. 8, as expected. However, the model revealed an unrealistic material response in constant width deformation (i.e. deformation stress much higher than in equal biaxial and the stress heads towards infinity in constant width deformation).

4.1.2 Mooney-Rivlin Model

Figures 9 and 10 depict the fitted results of the 2nd order Mooney-Rivlin model to PP and HIPS, respectively.

It can be seen that with the use of material parameters fitted from EB test data, the 2nd order *Mooney-Rivlin* model show a good agreement (apart from the pronounced yielding and strain softening exhibited by PP) between model

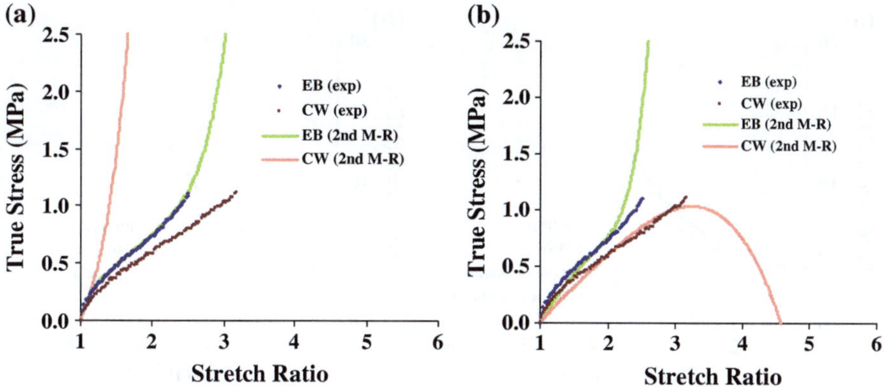

Fig. 10 Model prediction (2nd order M-R) to EB and CW responses of HIPS. **a** Fitted to equal biaxial test data. **b** Fitted to constant width test data

predicted EB response and the experimental results of both PP and HIPS, however, its reveals an unrealistic material response in constant width deformation. Employing the material parameters fitted from CW test data, the 2nd order *Mooney-Rivlin* model again shows a non-physical result for PP where the model revealed a dramatic decrease in tensile stress (towards negative) with the increase in stretch ratio above 2 (in EB) and 3.5 (in CW).

4.2 Physical Network Hyperelastic Model

4.2.1 neo-Hookean Model

Figures 11 and 12 show the fitted results of the neo-Hookean model to PP and HIPS, respectively. It was found that the *neo-Hookean* (*Gaussian* network) model is incapable of capturing the EB and CW deformation behaviour of both PP and HIPS used in this study. In addition, the model simulated EB and CW true stress-stretch ratio curves are observed to be nearly identical.

4.2.2 Three-Chain Model

The 4th-term extension of the inverse *Langevin* function was employed in the evaluation of the true stress-stretch ratio relationship of the 3-*chain* model. Figures 13 and 14 show the fitted results of the 3-chain model to PP and HIPS, respectively. It can be observed that the *non-Gaussian* 3-*chain* model is incapable of capturing the equal biaxial and constant width deformation of both PP and HIPS used in this study. In addition, the model prediction was found to be very similar to the *neo-Hookean Gaussian* network model.

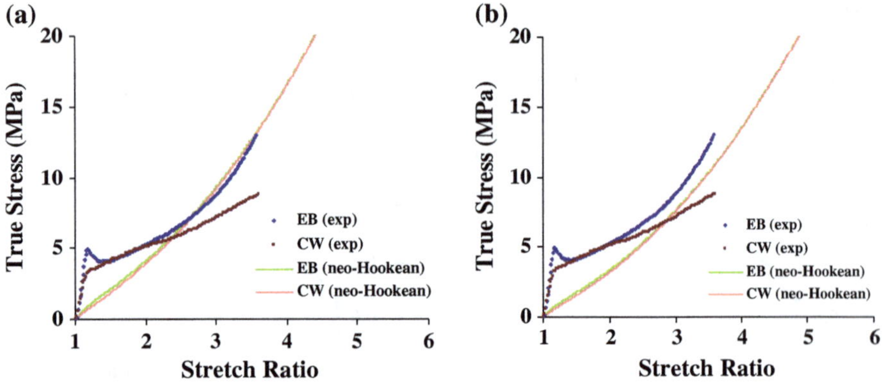

Fig. 11 Model prediction (neo-Hookean) to EB and CW responses of PP. **a** Fitted to equal biaxial test data. **b** Fitted to constant width test data

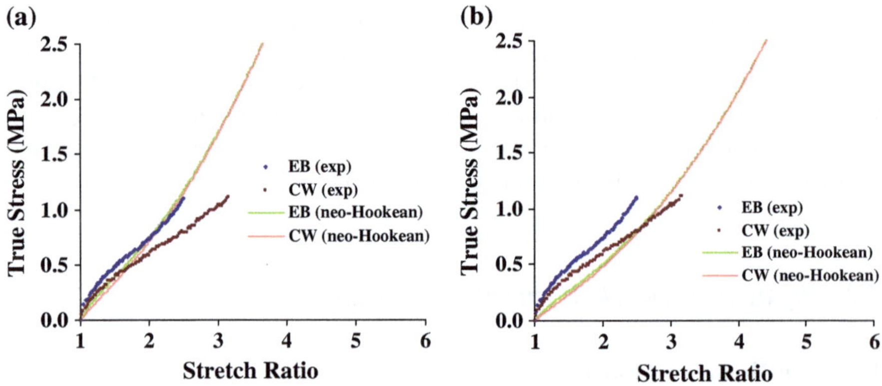

Fig. 12 Model prediction (neo-Hookean) to EB and CW responses of HIPS. **a** Fitted to equal biaxial test data. **b** Fitted to constant width test data

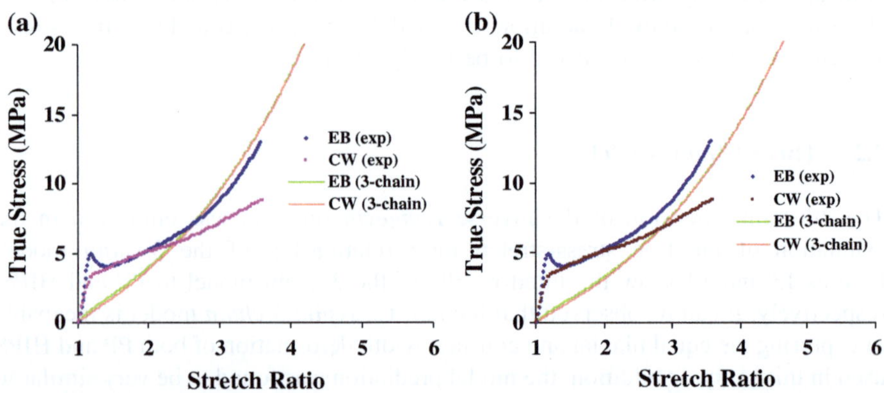

Fig. 13 Model prediction (3-chain model) to EB and CW responses of PP. **a** Fitted to equal biaxial test data. **b** Fitted to constant width test data

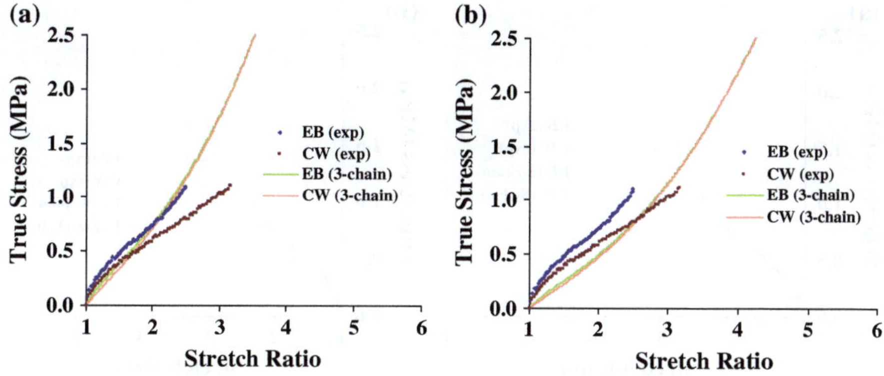

Fig. 14 Model prediction (3-chain model) to EB and CW responses of HIPS. **a** Fitted to equal biaxial test data. **b** Fitted to constant width test data

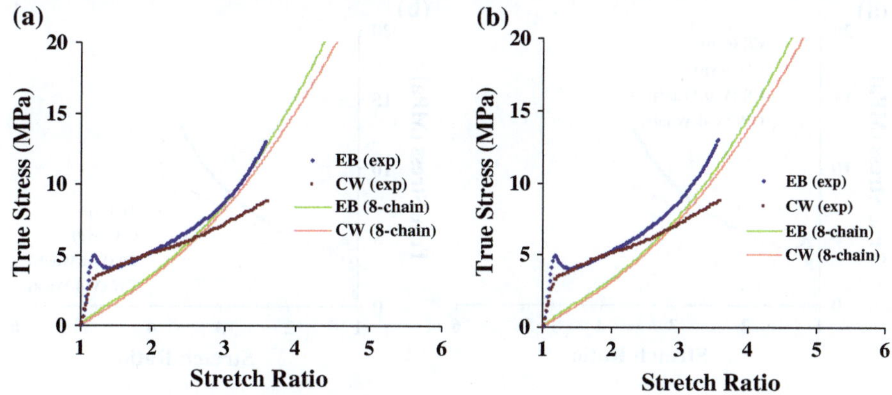

Fig. 15 Model prediction (8-chain model) to EB and CW responses of PP. **a** Fitted to equal biaxial test data. **b** Fitted to constant width test data

4.2.3 Eight-Chain Model

Similar to the prediction of the *neo-Hookean* (*Gaussian* network) and 3-*chain* (*non-Gaussian* network) models, the *non-Gaussian* 8-*chain* model was found incapable of accurately simulating the EB and CW deformation behaviour of both PP and HIPS used in this study, as shown in Figs. 15 and 16. The 4th-term extension of the inverse *Langevin* function was employed in the evaluation of the true stress-stretch ratio relationship of the 8-*chain* model.

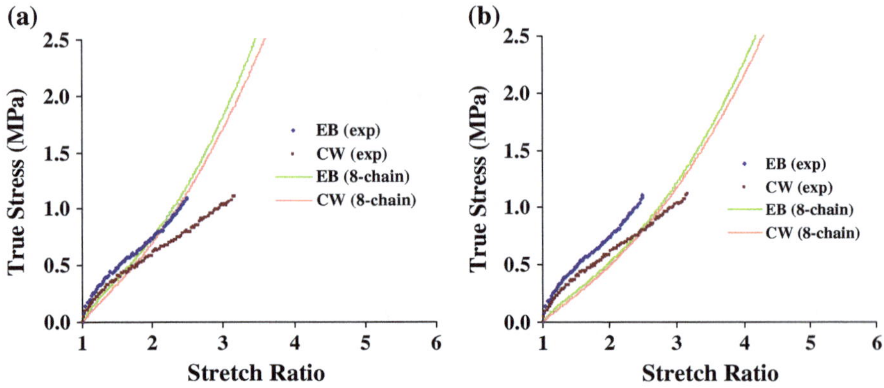

Fig. 16 Model prediction (8-chain model) to EB and CW responses of HIPS. **a** Fitted to equal biaxial test data. **b** Fitted to constant width test data

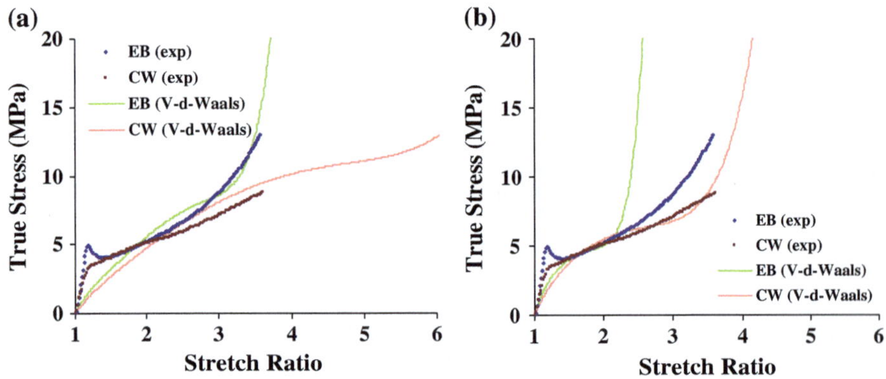

Fig. 17 Model prediction (Van der Waals) to EB and CW responses of PP. **a** Fitted to equal biaxial test data. **b** Fitted to constant width test data

4.2.4 Van der Waals Model

Figures 17 and 18 show the fitted results of the Van der Waals model to PP and HIPS, respectively. It can be observed that although the *Van der Waals* model gives a better representation of the deformation behaviour of PP compared to the *Gaussian* (*neo-Hookean*) and *non-Gaussian* (3-*chain*, 8-*chain*) network model, the accuracy of the model prediction is relatively poor. The simulated EB stress moves towards infinity at a stretch ratio found to be too low for finite element simulation of typical deep draw free surface moulding processes. On the other hand, the simulated true stress-stretch ratio curves of the *Van der Waals* model were found to be comparable to the experimental data of HIPS. In all cases, the least-square fit

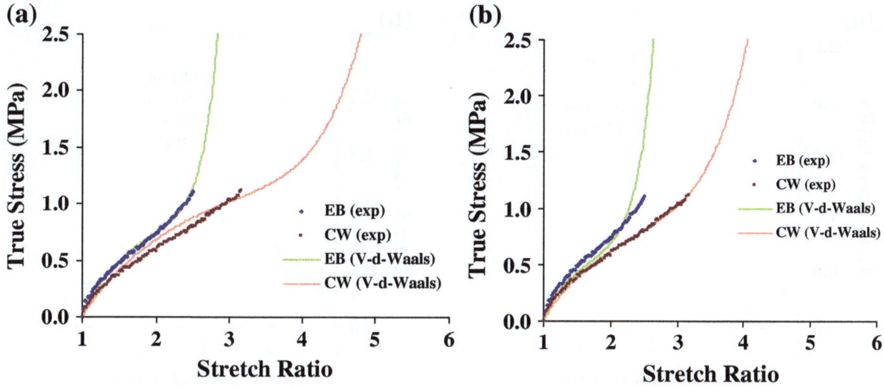

Fig. 18 Model prediction (Van der Waals) to EB and CW responses of HIPS. **a** Fitted to equal biaxial test data. **b** Fitted to constant width test data

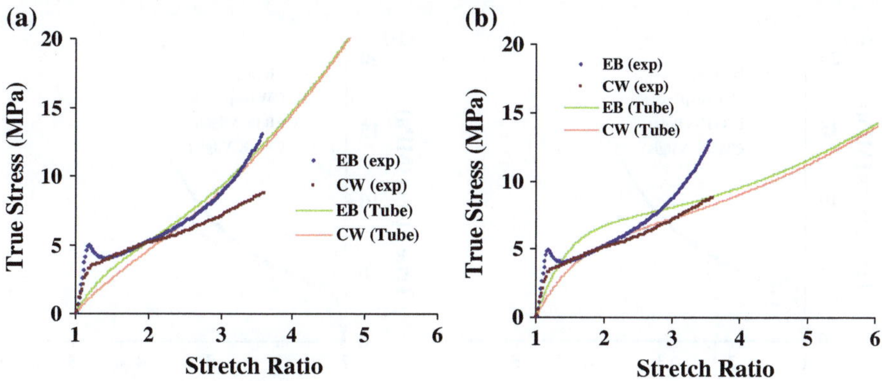

Fig. 19 Model prediction (Ball 'tube'model) to EB and CW responses of PP. **a** Fitted to equal biaxial test data. **b** Fitted to constant width test data

of the *Van der Waals* model parameters reveals that the simulated true stress-stretch ratio curve is very sensitive to slight changes in each of its material parameters, especially parameter β.

4.2.5 Ball "Tube" Model

From the nonlinear least-square fit results of *Ball* model, it can be observed that the model was unsuccessful in simulating the deformation behaviour of both PP and HIPS used in this study, as shown in Figs. 19 and 20 respectively. For PP, the simulated curves were virtually similar for both EB and CW deformation. For HIPS, it was observed that the optimal least-square fitted parameters produced

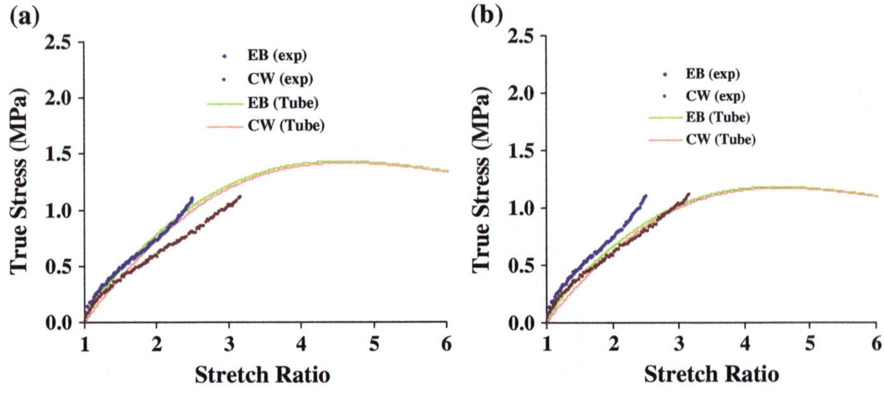

Fig. 20 Model prediction (Ball 'tube' model) to EB and CW responses of HIPS. **a** Fitted to equal biaxial test data. **b** Fitted to constant width test data

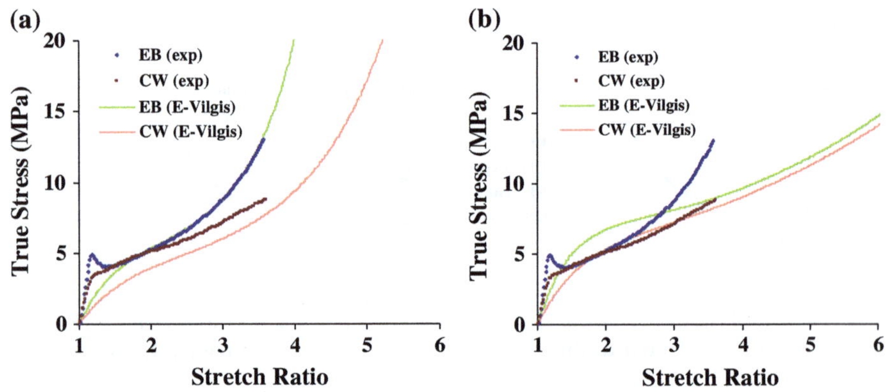

Fig. 21 Model prediction (Edwards-Vilgis) to EB and CW responses of PP. **a** Fitted to equal biaxial test data. **b** Fitted to constant width test data

stresses, which tend to drop with increasing stretch ratio, indicating the onset of the model's instability at large stretch.

4.2.6 Edwards-Vilgis Crosslink-Sliplink Model

Figures 21 and 22 show the fitted results of the Edwards-Vilgis model to PP and HIPS, respectively.

When material constants best fitted to experimental EB data are used, the *Edwards-Vilgis* model was found capable of accurately simulating the EB deformation responses of both PP (apart from the significant yielding found in the test

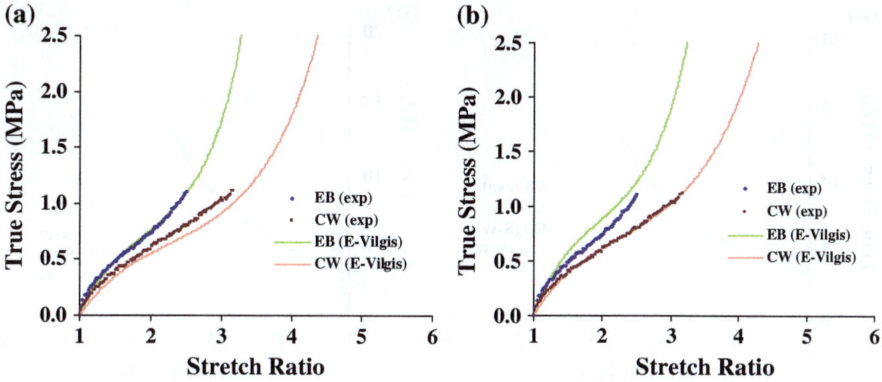

Fig. 22 Model prediction (Edwards-Vilgis) to EB and CW responses of HIPS. **a** Fitted to equal biaxial test data. **b** Fitted to constant width test data

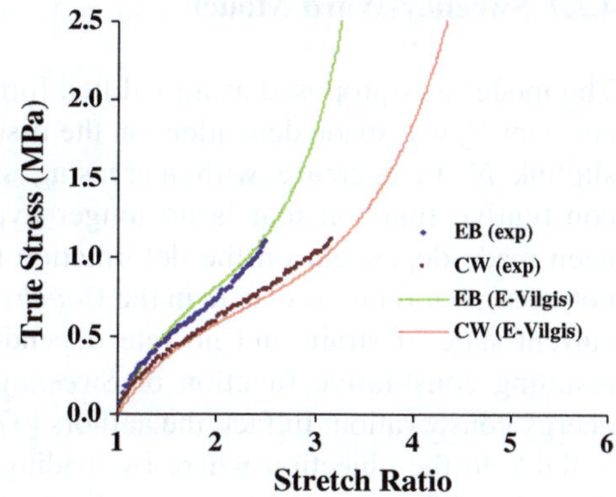

Fig. 23 Model prediction (Edwards-Vilgis) to EB and CW responses of HIPS. (Averaging material constant fitted from EB and CW)

data) and HIPS. However, the model under-predicted the true stress in CW deformation of PP. In the case of HIPS, only a slight discrepancy was found in the simulated constant width curve. When material constants best fitted to experimental CW data are used, the model not only gives unacceptable prediction in the equal biaxial curve, but also under predicts both the initial modulus and strain hardening in the constant width curve.

The slight discrepancy observed in the model predicted response of HIPS may be attributed to the minor variation in experimental data. Alternatively, an average of the characterised constants from both EB and CW test data of HIPS can be evaluated to give higher accuracy in the simulated responses, as shown in Fig. 23.

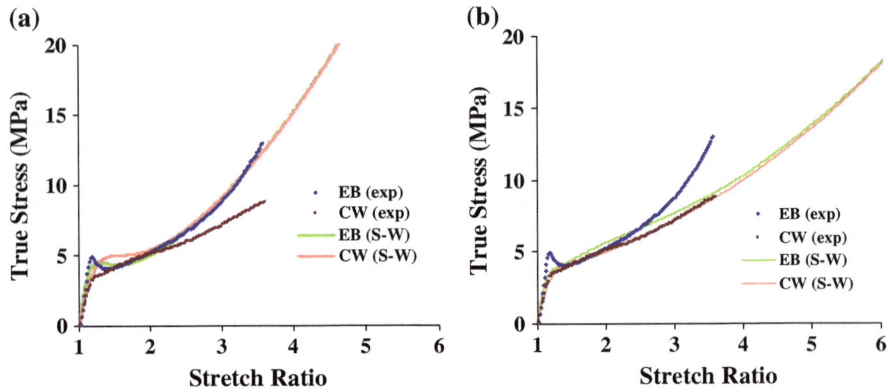

Fig. 24 Model prediction (Sweeney-Ward) to EB and CW responses of PP. **a** Fitted to equal biaxial test data. **b** Fitted to constant width test data

4.2.7 Sweeney-Ward Model

The model was proposed as a modified form of the *Ball* model, where the sliplink constant N_s was made dependent on the first strain invariant to allow the number of sliplink N_s to decrease with increasing strain. This modification resulted in a constitutive function that is no longer hyperelastic (since the strain energy has been made dependent on the deformation path and a strain energy function does not exist), but remains elastic in the *Cauchy* sense (since stress depends only on the current state of strain and no rate dependency is accounted for). Therefore, the resulting constitutive function of *Sweeney-Ward* no longer obeys the theory of energy conservation. In fact, the authors [37] did point out that the proposed model is liable to the objection where by loading via a deformation path and unloading through a different one, the material model may acquire more work on unloading than was required to load it. The cause of the problem was attributed to the implicit assumption that the sliplink number N_s will essentially recover to its original value on unloading, however in practice, the deformation is expected to have permanent effects.

The entire experimental equal biaxial true stress-stretch ratio data of PP, including the yield and strain softening were used in the least-square fit procedure for the *Sweeney-Ward* model. Figures 24 and 25 show the fitted results of the *Sweeney-Ward* model to PP and HIPS, respectively. When material constants best fitted to EB test data are used, it can be observed that the model is capable of accurately representing the equal biaxial deformation response of both PP and HIPS. However, the model simulated CW response was found to be virtually identical to that of the EB deformation. A similar trend is observed when the model is fed with material constants characterised from constant width test data.

It should be noted that in the nonlinear least-square fitting procedure of the *Ball*, *Edwards-Vilgis* and *Sweeney-Ward* models, the material constant corresponding to

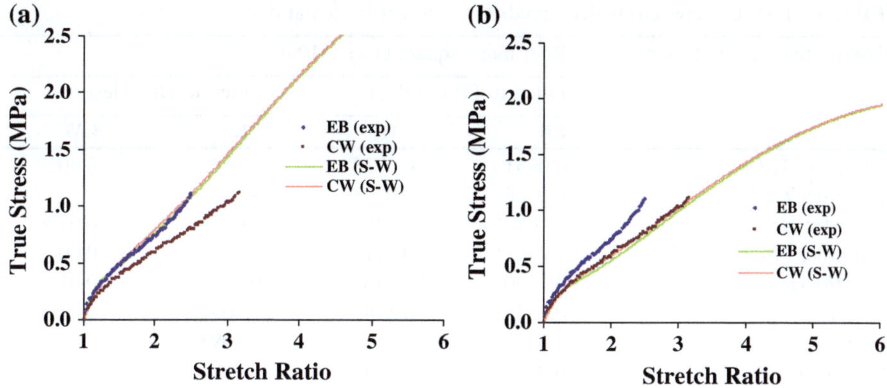

Fig. 25 Model prediction (Sweeney-Ward) to EB and CW responses of HIPS. **a** Fitted to equal biaxial test data. **b** Fitted to constant width test data

the number of crosslink N_c was assigned to 0 for HIPS, due to the fact that HIPS is an amorphous non-crosslinked material and its modulus is assumed to be contributed by chain entanglements or sliplinks alone. For the PP, the appearance of the crystalline phase was assumed to act as artificial crosslinks within the material. This can be attributed to the fact that at forming temperature, the mobility of the amorphous phase is much higher than the crystalline phase, due to the fact that the solid phase forming temperature of PP is far higher than its glass transition temperature (T_g at ≈ -5 °C as determined from DMTA) and thus the modulus is assumed to be contributed from both crosslink and sliplink.

Quantitative fit of the hyperelastic or *Cauchy's* elastic models investigated in this work can be visualised by evaluating the 'root mean square error' between the model's prediction and the experimental data through

$$RMSE = \left[\frac{1}{N} \cdot \sqrt{\sum_{i=1}^{N} \left(\sigma_i^{test} - \sigma_i^{model}\right)^2}\right] \tag{37}$$

where N is the number of experimental true stress-stretch ratio data pairs, σ_i^{test} is the true stress value from test data and σ_i^{model} is the model generated true stress.

When fitted to either experimental EB or CW data, the 'root mean square error' of the models' prediction can be shown in Table 1.

It should be noted that while the values of the 'root mean square error' represents the goodness of fit to either experimental EB or CW data, they do not represent the quantitative fit of the models to both deformation modes (e.g. a material model might accurately capture EB deformation but poorly predict the CW response).

Table 1 'RMSE' between models' prediction and experimental data

Constitutive material model	Root mean square error (MPa)			
	Fitted to PP test data		Fitted to HIPS test data	
	EB	CW	EB	CW
1st order *Ogden*	0.0951	0.0728	0.0045	0.0029
2nd order *Ogden*	0.0306	0.0728	–	–
1st order *Mooney-Rivlin*	0.1080	0.1481	0.0047	0.0106
2nd order *Mooney-Rivlin*	0.0402	0.0919	0.0016	0.0047
Neo-Hookean	0.1260	0.1481	0.0086	0.0106
3-*chain*	0.1333	0.1545	0.0092	0.0114
8-*chain*	0.1410	0.1513	0.0095	0.0110
Van der Waals	0.0920	0.0449	0.0026	0.0017
Ball 'tube'	0.0879	0.0508	0.0045	0.0037
Edwards-Vilgis	0.0744	0.0507	0.0033	0.0017
Sweeney-Ward	0.0252	0.0101	0.0017	0.0011

5 Conclusions

From the analyses carried out on a range of constitutive hyperelastic models (both phenomenological and physical), it is observed that the simulated true stress-stretch ratio curves from the *Ogden*, *Van der Waals* and *Edwards-Vilgis* models are in reasonably good agreement with the experimental EB and CW deformation behaviour of amorphous HIPS. However, the material models considered in this work were generally incapable of accurately capturing the deformation behaviour of PP, especially in terms of simulating the initial *Young's* modulus, yield and strain softening. In this respect, further development would be required to develop a new constitutive material model that can accurately capture the complex deformation behaviour of PP.

References

1. Hooke, R., De Potentia Restitutiva or of Spring Explaining the Power of Springing Bodies, p. 23. London (1678)
2. Markovitz, H.: The emergency of rheology. Physics Today (American Institute of Physics) **21**(4), 23–33 (1968)
3. Truesdell, C.A., Cauchy's First Attempt at Molecular Theory of Elasticity, Bollettino di Storia delle Scienze Matematiche. Il Giardino di Archimede **1**(2), 133–143 (1981)
4. Bogolyubov, A.N., Augustin Cauchy and His Contribution to Mechanics and Physics (Russian), Studies in the History of Physics and Mechanics, pp. 179–201. Nauka, Moscow (1988)
5. Dahan-Dalmédico, A., La Propagation Des Ondes En Eau Profonde Et Ses Développements Mathématiques (Poisson, Cauchy, 1815–1825), in The History of Modern Mathematics II, pp. 129–168. Boston, MA (1989)

6. Truesdell, C.A.: Cauchy and the modern mechanics of continua. Rev. Hist. Sci. **45**(1), 5–24 (1992)
7. Mooney, M.: A theory of large elastic deformation. J. Appl. Phys. **11**(9), 582–592 (1940)
8. Rivlin, R.S.: Large elastic deformations of isotropic materials, I, II, III, fundamental concepts. Philos. Trans. R. Soc. Lond. Ser. A **240**, 459–525 (1948)
9. Rivlin, R.S.: Large elastic deformations of isotropic materials, IV, further developments of the general theory. Philos. Trans. R. Soc. Lond. Ser. A **241**(835), 375–397 (1948)
10. Ogden, R.W., Large deformation isotropic elasticity: on the correlation of theory and experiment for incompressible rubberlike solids. In: Proceedings of the Royal Society of London, Series A, Mathematical and Physical Sciences, vol. 365, pp. 565–584 (1972)
11. Treloar, L.R.G.: Stress-strain data for vulcanized rubber under various types of deformation. Trans. Faraday Soc. **40**, 59–70 (1944)
12. Ward, I.M., Mechanical Properties of Solid Polymers, 2nd edn. Wiley, Hoboken (1983)
13. Treloar, L.R.G.: The elasticity of a network of long chain molecules I. Trans. Faraday Soc. **39**, 36–64 (1943)
14. Treloar, L.R.G.: The elasticity of a network of long chain molecules II. Trans. Faraday Soc. **39**, 241–246 (1943)
15. Kuhn, W., Grun, F.: Beziehungen zwichen elastischen Konstanten und Dehnungsdoppelbrechung hochelastischer stoffe. Kolloideitschrift **101**, 248–271 (1942)
16. James, H.M., Guth, E.: Theory of elastic properties of rubber. J. Chem. Phys. **11**(10), 455–481 (1943)
17. Wang, M.C., Guth, E.: Statistical theory of networks of non-gaussian flexible chains. J. Chem. Phys. **20**, 1144–1157 (1952)
18. Arruda, E.M., Boyce, M.C.: A three-dimensional constitutive model for the large stretch behavior of rubber elastic materials. J. Mech. Phys. Solids **41**(2), 389–412 (1993)
19. Treloar, L.R.G., Riding, G., A non-gaussian theory for rubber in biaxial strain I, mechanical properties. In: Proceedings of the Royal Society of London, Series A, Mathematical and Physical Sciences, vol. 369, pp. 261–280 (1979)
20. Wu, P.D., Van der Giessen, E.: On improved 3-D non-Gaussian network models for rubber elasticity. Mech. Res. Commun. **19**(5), 427–433 (1992)
21. Wu, P.D., Van der Giessen, E.: On improved network models for rubber elasticity and their applications to orientation hardening in glassy polymers. J. Mech. Phys. Solids **41**(3), 427–456 (1993)
22. Flory, P.J., Erman, B.: Theory of elasticity of polymer networks. Macromolecules **15**(3), 800–806 (1982)
23. Ball, R.C., Doi, M., Edwards, S.F., Warner, M.: Elasticity of entangled networks. Polymer **22**(8), 1010–1018 (1981)
24. Doi, M., Edwards, S.F.: The theory of polymer dynamics. Oxford University Press, Oxford (1986)
25. Kilian, H.G.: Equation of state of real networks. Polymer **22**, 209–217 (1981)
26. Kilian, H.G.: Energy balance in networks simply elongated at constant temperature. Colloid Polym. Sci. **259**, 1084–1091 (1981)
27. Vilgis, T.H., Kilian, H.G.: The van der waals-network—a phenomenological approach to dense networks. Polymer **25**, 71–74 (1984)
28. Kilian, H.G., Vilgis, T.H.: Fundamental aspects of rubber-elasticity in real networks. Colloid Polym. Sci. **262**, 15–21 (1984)
29. Kilian, H.G.: An interpretation of the strain-invariants in largely strained networks. Colloid Polym. Sci. **263**, 30–34 (1985)
30. Edwards, S.F., Vilgis, T.H.: The effect of entanglements in rubber elasticity. Polymer **27**(4), 483–492 (1986)
31. Edwards, S.F., Vilgis, T.H.: The stress-strain relationship in polymer glasses. Polymer **28**(3), 375–378 (1987)
32. Rouse, P.E.: A theory of the linear viscoelastic properties of dilute solutions of coiling polymers. J. Chem. Phys. **21**, 1272–1280 (1953)

33. Flory, P.J.: Principles of Polymer Chemistry. Cornell University Press, New York (1966)
34. Flory, P.J.: Theory of Polymer Networks. The Effect of Local Constraints on Junctions. J. Chem. Phys. **66**(12), 5720 (1977)
35. Sweeney, J., Ward, I.M.: The modeling of multiaxial necking in polypropylene using a sliplink-crosslink theory. J. Rheol. **39**(5), 861–872 (1995)
36. Sweeney, J., Kakadjian, S., Craggs, G., Ward, I.M., The multiaxial stretching of polypropylene at high temperatures. In: ASME Mechanics of Plastics and Plastic Composites, MD vol. 68 / AMD vol. 215 pp. 357–364 (1995)
37. Sweeney, J., Ward, I.M.: A constitutive law for large deformations of polymers at high temperatures. J. Mech. Phys. Solids **44**(7), 1033–1049 (1996)

MIX
Papier aus verantwortungsvollen Quellen
Paper from responsible sources
FSC® C105338

If you have any concerns about our products,
you can contact us on
ProductSafety@springernature.com

In case Publisher is established outside the EU,
the EU authorized representative is:
Springer Nature Customer Service Center GmbH
Europaplatz 3, 69115 Heidelberg, Germany

Printed by Libri Plureos GmbH
in Hamburg, Germany